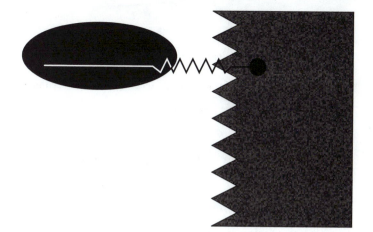

Becoming an
Electronics
Technician

Securing Yo

RONALD A. REIS
Los Angeles Valley College

Prentice
Hall

Upper Saddle River, New Jersey
Columbus, Ohio

Library of Congress Cataloging-in-Publication Data

Reis, Ronald A.
   Becoming an electronics technician : securing your high-tech future / Ronald A. Reis.—4th ed.
      p. cm.
     Includes index.
     ISBN 0-13-093219-1 (alk. paper)
      1. Electronics—Vocational guidance. 2. Electronic technicians. I. Title.
  TK7845 .R4 2002
  621.381′023—dc21

                                              2001026710

Editor in Chief: Stephen Helba
Assistant Vice President and Publisher: Charles E. Stewart, Jr.
Assistant Editor: Delia K. Uherec
Production Editor: Alexandrina Benedicto Wolf
Design Coordinator: Diane Ernsberger
Cover Designer: Thomas Borah
Production Manager: Matthew Ottenweller

This book was set in Dutch 823 by The Clarinda Company. It was printed and bound by R. R. Donnelley & Sons Company. The cover was printed by Phoenix Color Corp.

Pearson Education Ltd., *London*
Pearson Education Australia Pty, Limited, *Sydney*
Pearson Education Singapore, Pte. Ltd
Pearson Education North Asia Ltd., *Hong Kong*
Pearson Education Canada, Ltd., *Toronto*
Pearson Educación de Mexico, S.A. de C.V.
Pearson Education — Japan, *Tokyo*
Pearson Education Malaysia, Pte. Ltd.
Pearson Education, *Upper Saddle River, New Jersey*

Prentice
Hall

10 9 8 7 6 5 4 3 2 1
ISBN 0-13-093219-1

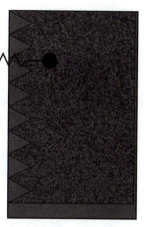

Preface

As I travel about, talk with electronics instructors, and visit schools, I detect—particularly among first-semester electronics students—confusion concerning the electronics world and their place in it. Many students, while already enrolled in electronics programs, still do not have a clear appreciation of what it is like to be or to become an electronics technician. This is unfortunate. If students are to do well in their studies, greeting each school day with a "plan and a purpose," they need to better understand the world of electronics, the role they will eventually play in it, and the type of preparation needed for success as as electronics technician. This book is an attempt to give electronics students the encouragement and support they need to become part of the exciting future in electronics.

Electronics students have many questions that need to be answered. They want to know about the electronics industry—its strength and its diversity. They are curious about the people who populate this industry, their occupations, and their careers. Students are particularly interested in the role of the electronics technician in today's and tomorrow's workplace. And they want to meet electronics technicians, to hear from them, in their own words, what it is like in the real world.

Students also want to know what it takes to become educated and trained in electronics: what type of schooling is available and what is to be learned. They want help in succeeding as electronics students: planning their day and developing their study skills. Furthermore, they know electronics can be more than just books and theory. They want to know what it takes to be an "electronics activist." And they want clarification on the myriad of information technology (IT) certifications, from A+ certification to advanced networking credentials. Finally, of course, students want to start a career as an electronics technician. They want to know how to find, get, and keep their first job in the field.

In sum, electronics students want to know what it means to be an electronics technician and what it takes to become such a technician. I hope this book will provide some answers to their questions.

But they also want more. Particularly, at this stage in their studies, students are eager to "do" electronics. They want to get their hands on the "goodies": components, breadboards, and test equipment. They want to "make like" an electronics technician. In Part III, we give them the opportunity.

Where can such a book be used? Actually, anywhere in the electronics curriculum; I believe no student should graduate without having been exposed to this material. Of course, it makes sense to read such a book at the "front end" of the program, in an electronics survey class or an introduction to dc/ac fundamentals. I know time is limited, and instructors are hard pressed to present what is already in their course outline. But this book is short and, I am told, easy to read. More to the point, a dc/ac fundamentals course, which tends to be quite theoretical, even esoteric, is just the place where a supplemental book of this type can liven up the subject, while giving students direction and support for their study tasks ahead.

Just reading the book, however, is not enough. Students must become involved in the subjects presented. To that end, each chapter in Parts I and II contains individual and group activities, along with issues for class discussion, to involve students and engage the entire class in meaningful dialogue. Take full advantage of the material presented.

This edition contains a number of additions and changes:

- Data (charts, diagrams, and supporting text) have been revised throughout.
- Information on job shadowing has been included (Chapter 5).
- Information on ISCET certification has been expanded (Chapter 7).
- A completely new chapter (Chapter 8) on information technology (IT) certification, with emphasis on A+ certification, has been added.
- Material on online job search techniques has been expanded (Chapter 9).

- Information on negotiating salary and benefits has been included (Chapter 9).

- A completely new chapter (Chapter 16) on virtual electronics has been added.

- Five new projects have been added to Chapter 17.

- Appendix 1, has been completely updated and now contains 34 listings.

- A glossary of over 50 terms has been added.

If you have any questions or suggestions, please contact me via e-mail at Ronelect@aol.com.

Acknowledgments

Once again, thanks to Charles Stewart, my acquisitions editor, for his enthusiasm and support. I also wish to thank Alex Wolf, who saw the book through to completion.

Ronald A. Reis

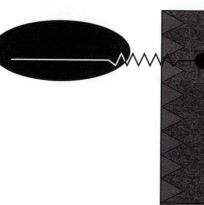

Contents

CHAPTER 6

Succeeding as an Electronics Student: Making the Most of Your School Day   129

CHAPTER 7

Being an "Electronics Activist": Electronics Beyond the Classroom   151

CHAPTER 17

Useful Electronic Projects You Can Build   **301**

APPENDIX 1

Electronic Kit and Part Sources   **329**

APPENDIX 2

The Big Three Electronics Magazines   **337**

Glossary   **339**

Index   **345**

Becoming an Electronics Technician

Securing Your High-Tech Future

PART

1

Being an Electronics Technician

In Part I we explore what it's like to work as an electronics technician.

In Chapter 1, "The Electronics Industry: In the Air and Everywhere," we examine the industry as a whole, its diversity and success. We take a brief trip back in time to see how electronics began and what major developments fueled its growth. We go on to analyze its organization and classification. And we take an upbeat look at electronics, surveying new products and services that are bursting forth in the first years of the new millennium.

Then it's on to Chapter 2, "Electronics: Field of Dreams," where we discover the occupations that populate this enormous industry. We concentrate on the careers of an electronics engineer, an electronics technologist, and an electronics technician. We examine job growth trends as related to electronics. And we conclude with an inquiry into the workplace revolution by exploring issues such as the *knowledge gap* and *workplace technical literacy.*

In Chapter 3, "The Electronics Technician: Bringing Electronics to Life," we zero in on the career of the electronics technician. We begin by developing a four-part classification for electronics technicians and examining it in detail. We then go on to a scrutiny of the "hard-side" and "soft-side" skills beyond electronics that every technician needs to develop. Here we emphasize mechanical know-how and computer know-how, as well as communication, selling, training, and supervisory skills. Technical knowledge alone is not enough for today's electronics technician.

Finally, in Chapter 4, "Electronics Technician Profiles: Meet Ron, David, Kay, Gideon, and Eric," we present five in-depth technician interviews to hear what it's like out there on the job. Ron is an industrial electronics technician working for a large private corporation. Kay is a communications technician employed with the Southern California Rapid Transit District. Gideon is a field service representative for a well-known office machine company, Rico. David is an A+ certified computer tech in a rewarding management position. And Eric owns his own business in the consumer electronics field.

When you have completed Part I, "Being an Electronics Technician," you'll know about the electronics industry, the people who work in it, and the role of the electronics technician, and you will have met five bright, hard-working techs. You will then be ready for Part II, "Becoming an Electronics Technician."

The Electronics Industry:
In the Air and Everywhere

Objectives

In this chapter you will learn:

- About the electronics industry's staggering growth.
- How electronics affects us, particularly in consumer electronics.
- How an $8 billion, 288-satellite project, Teledesic, will transform telecommunications.
- How electronics began with the lowly vacuum tube in 1906.
- How the average North American encounters 300 microcomputers in a single day.
- How the electronics field is organized and classified.
- How Moore's Law is doubling computer power every 18 months.
- About new electronic products and services coming online in the first decade of the new millennium.

"Doctors were off the golf course and near a phone. Real estate agents were adrift in a red-hot market. Parents lost the lifeline anchoring them to children growing up in a tougher, scarier world. And drug dealers suffered a rare business slump." Thus, Anne-Marie O'Connor, writing in the Los Angeles Times, gives us her take on the great 1998 pager blackout. On May 19, a single PanAmSat satellite, Galaxy 4, hovering 22,300 miles above Earth spun out of orbit, and America's estimated 49.5 million pagers went silent. If anyone needed proof of our reliance on electronics, the Galaxy satellite knockout cinched it. As one user declared, "Take away my pager and I'd be back in the Middle Ages, I can't live without one."

Today, pagers are even more a part of many lives, although the tremendous growth in wireless phones, with their web-connect features, are slowing growth somewhat. But pagers aren't just about numbers anymore. And, with two-way advanced services that let consumers send messages from one pager to another, or to a telephone, email address, or fax machine, the $4-billion pager industry is definitely an on-going player in the wireless communications market. In 1990 there were 10 million pager users in the United States, but by 2000 the number had climbed to an astonishing 60 million.

In Chapter 1, we examine the industry responsible for this pager dependency. We begin by exploring its mammoth success, with particular emphasis on the consumer electronics sector touching us directly. Next, we look back to see who and what have made electronics so pervasive, so omnipresent. Third, we penetrate the industry, attempting to identify and clarify its many subfields. Finally, we probe ahead, striving to get a feel for the huge product range and career opportunities electronics promises as the first years of the 21st century unfold.

Diversity and Success

The electronics industry, in the broadest definition, provides electricity-based products and services in increasing number and variety in order to enhance our work, play, learning, health, safety, and security—in other words, to provide a higher standard of living.

All these electronic products and services have one thing in common: They contain electronic components, connected to form electronic circuits, that in turn control the flow of current, or electricity. Whether the product is a biometric iris scanner checking the eye's complexity, a 128-megabyte digital music player, film-quality digital TV, an ultrasound detector painlessly examining your stomach for an ulcer, or a missile being guided to its target, each manipulates and controls the movement of free electrons within electronic circuits at close to the speed of light. The result?—wonders ranging from frivolous games to life-saving necessities.

The industry responsible for all this has grown at a staggering rate— 23.8 percent from 1994 to 1998. Today, electronics is the world's largest economic enterprise.

According to a Commerce Department report titled "The Emerging Digital Economy" released on April 15, 1998, "The high-tech industry today accounts for more than 8% of the national output of goods and services—more than $600 billion a year—with computer and communication sectors growing twice as fast as the economy as a whole."

Moreover, high-tech companies aren't the only ones involved. "Among the industries investing most heavily in information technology," the report continues, "are real estate, radio and television broadcasting, auto repair services, and motion picture production."

Today, nationwide, investments in information technology account for 45 percent of all business equipment investment. In the 1960s, the figure was 3 percent.

Moreover, all these ventures are creating well-paying jobs. Computers and the Internet have dramatically transformed the nation's economy in the last few years, creating over 7 million high-paying jobs. In early 2001, anywhere from 220,000 to 843,000 information technology (IT) jobs went unfilled (see Chapter 8).

In 1998, according to the above cited Commerce Department report, the average information technology job paid $46,000 a year. That compares with $28,000 for the private sector as a whole. Clearly, a career in technology is a career to secure your high-tech future.

To see how electronics affects us in just one subfield, look at what's happening in consumer electronics. As shown in Figure 1.1, the penetration of selected consumer electronics products into U.S. households is staggering. Virtually every home had a color TV and nine out of ten had a VCR. Today, computers are in half of all residences, 60 million households.

The Consumer Electronics Association estimates that more consumer electronic products will be launched from 1998 to 2003 than during the entire previous history of the industry. Indeed, if you attended the 2001 Consumer Electronics Show in Las Vegas, you saw:

- The Xbox, Microsoft's long-awaited and ultra-cool new video game console with enhanced graphics, increased memory, and greater multi-user capabilities.
- Plug-ins that turn your Palm into a telephone, camera, or music player.

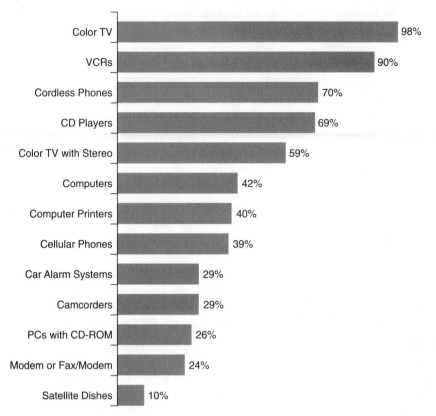

Figure 1.1
U.S. household penetration of selected consumer electronics products January, 1998.
Source: Consumer Electronics Manufacturers Association (CEMA).

- Home networks that control alarm systems, distribute TV, and connect to the Internet.
- Machines that recommend TV shows based on your viewing preferences.
- Radios that give you better sound, enhanced satellite program options, and the ability to get the signal anywhere in the U.S.
- Home entertainment systems with Internet connections that download DVD videos and MP3 music files.
- E-wear eyeglasses with computer screens.
- Digital cameras in wrist watches.
- Digital music players the size of a matchbox.

- Video game consoles that play games, DVDs, and CDs.
- Mobile phones with miniaturized laptops and headsets.
- Mega data holders in formats the size of a stamp.
- Television sets with Internet browsing for email, movies, music, and shopping.
- Flash memory disks that can be used with a PC, portable player, or home stereo system.
- Personal Digital Assistants (PDSs) that use voice recognition to receive commands and data.

Of course, whether any of these "thingies" will make our lives better— or just more cluttered and ultimately confusing—is anyone's guess.

And with electronics, we're all fortunate—as time goes on, you get more for less. Figure 1.2 shows the changes in consumer electronics prices compared to all consumer products from 1991 to 1997. As the price for most everything else goes up, for consumer electronic products it's the other way around.

Of course, all this cost saving on the consumer end requires big-buck investments on the design and manufacturing side. Teledesic, based in Kirkland, Washington, is building a 288-satellite project—expected to cost about $9 billion—that will connect computer networks and provide high-speed Internet access and videoconferencing worldwide.

Projects such as Teledesic are just an example of the U.S.-based electronic product and service innovations bursting forth in the new millennium. Although the United States is not the only player in the global electronics market, given its strength in military electronics, aerospace,

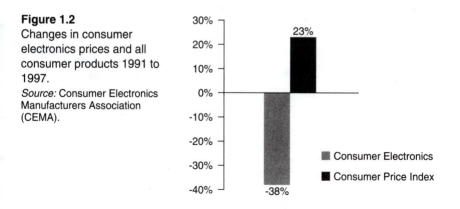

Figure 1.2
Changes in consumer electronics prices and all consumer products 1991 to 1997.
Source: Consumer Electronics Manufacturers Association (CEMA).

satellites, communications systems, and software, its future in the $2.6-trillion-a-year electronics industry looks bright indeed.

And, of course, there is the Internet, not exactly a project, more like a whole way of life. Internet usage is climbing steadily. It took radio 38 years to reach 50 million listeners. It took television 13 years to extend to the same number of viewers. The Internet did the same thing in just 4 years. Dot-com comings and goings aside, Internet usage in the U.S. topped 150 million by early 2001.

Just look at these Internet facts:

- By the end of 1999, 45 percent of all U.S. households were online. By 2005, 74 percent of all U.S. households will be online.
- In the United States, Internet commerce will reach $841.4 billion by 2003. Worldwide Internet commerce will reach $1.44 trillion by 2003.
- Ten million 16- to 22-year-olds account for $4.5 billion in online sales. A full 84 percent of 16- to 22-year-old shoppers use a parent's credit card when buying online.
- By 2002, 63 percent of all U.S. firms will be engaged in e-commerce.
- E-commerce comprised 0.8% of U.S. GDP in 1999. E-commerce will comprise 8.9% of U. S. GDP in 2003.
- Over 7 million Internet pages are coming on line daily.

TABLE 1.1
Average Internet U.S. User

Sessions per week	6
Unique sites visited	6
Pageviews per week	218
Pageviews per surfing session	37
Weekly surf time	3 hours, 6 minutes
Average session	31 minutes, 50 seconds
Duration of pageview	52 seconds
Average click rate for top banners	0.0035
Frequent Internet users	70,459,872
Total U.S. Internet Users	158,499,104

Data for week ending December 10, 2000
Source: "Nielsen//NetRatings," Business 2.0.

Table 1.1 shows characteristics of the average Internet user for the week ending December 10, 2000, according to "Nielsen//NetRatings," as reported in *Business 2.0.*

We'll have more to say about the Internet throughout the book. For electronics technicians, as for just about everyone, it has changed everything.

From Then Until Now

It all started—sort of—with Benjamin Franklin (1706–1790). Actually, Mr. Franklin's late-eighteenth-century experiments with a kite, a key, and lightning packing billions of watts of power did little to advance the science of electricity, while almost getting the inventor killed. Rather, it was the Italian Alessandro Volta (1745–1827), and his electric pile (battery), developed in 1801, that heralded the **electrical age** and a century of momentous developments. By the late nineteenth century, the United States and much of Europe were sending telegrams, talking on the telephone, working at night by electric light, traveling by electric trolley, and listening to recorded music on the gramophone. Electricity had become an influential part of many people's lives.

Electricity was a factor, but not electronics—not yet. It took Lee De-Forest (1873–1961) and his invention of the audion **vacuum tube** in 1906 to bring on the **age of electronics.**

The first major product to incorporate the audion (an amplifying component) was, of course, radio. But in its very early years, no one knew what to make of radio, what to do with it, or what, if anything, it was good for. In those days, individuals associated with radio thought in terms of private communications, as with cable, the telegraph, and the telephone. At the very least, radio would have to be confined to transmission from one point to another: point-to-point, or *narrowcast,* as it was called. Ship-to-shore communication was an obvious application. It wasn't until David Sarnoff's concept—the "Radio Music Box"—finally caught on in the early 1920s that point-to-mass, or *broadcast,* became a reality.

"I have in mind a plan of development," the future CEO of RCA began in a memorandum dated September 30, 1915, "that would make radio a household utility in the same sense as the piano or phonograph. The idea is to bring music into the house by wireless. . . . For example, a radiotelephone transmitter having a range of, say 25 to 50 miles can be installed at a fixed point where instrumental or vocal music or both are produced. . . . The receiver can be designed in the form of a "Radio Music Box" . . . and placed on a table in the parlor or living room, the switch set accordingly, and the transmitted music received" (as quoted in Kenneth Bilby, *The General: David Sarnoff and the Rise of the Communications Industry*).

A decade later, when radio advertising took hold, the Golden Age of Radio materialized. Hundreds of transmitters—and millions of radio receivers—were scattered across the land.

During the 1930s, while radio matured, experiments in an even more dramatic vacuum tube-based communications medium were forging ahead, most notably in the laboratories of RCA. Wireless communication of moving pictures, known as television, was ready for full-scale public demonstration at the 1939 New York World's Fair. It was an instant hit. But World War II brought further commercial development to a halt. It wasn't until the postwar period of the late 1940s and early 1950s that television truly established itself in U.S. living rooms.

While the war put a temporary damper on television, the global holocaust was actually a catalyst for another towering product of the vacuum-tube era, the digital computer. The first such "electronic brain," completed in 1946, was known as ENIAC, for *e*lectronic *n*umerical *i*ntegrator *a*nd *c*alculator. This behemoth contained 18,000 vacuum tubes and filled a room the size of a typical college lecture hall.

In 1947, the vacuum tube met its eventual replacement, the **transistor.** Far more reliable, much smaller, less expensive, and not nearly as power-hungry as its predecessor, the transistor revolutionized electronic product development. Clearly the most notable creation of the time was the ubiquitous, portable transistor radio of the 1960s and 1970s.

Rapid changes, however, were coming on fast, a defining characteristic of the industry by now. In the mid-1960s, the **integrated circuit,** or chip, as it is often called, began to find its way into commercial and military products. Integrated circuits (ICs) are more than just electronic components; rather, they are whole circuits on a tiny sliver of silicon. Today's ICs contain the equivalent of millions of transistor elements in a space less than the size of a fingernail. See Figure 1.3.

One of the earliest commercial products to incorporate a large number of ICs was the video game. Its striking success upon introduction was to be repeated again and again with a wide range of electronic products and services. Look what happened when the first coin-operated "pong" game hit the bars of Northern California in 1972 (Ronald Reis, *Understanding Electronics and Computer Technology*):

Figure 1.3
Integrated circuits.
Source: Copyright of Motorola,
Inc. Used by permission.

Credit for developing the video game usually goes to Nolan Bushnell, the founder of Atari. In 1972 he introduced a coin-operated pong game to patrons of Andy Capp's bar in Sunnyvale, California. As the story goes, Nolan was called in to check out the game's circuitry after the unit had stopped working. He discovered that the hundreds of TTL ICs (transistor-transistor-logic integrated circuits) were functioning just fine. But when he checked the coin box, it was jammed to capacity. The rest is history.

The pong game, however, is not what 1972 will be remembered for in electronics circles. In that year a far more significant event took place— the development of the first 8-bit **microprocessor.** Known as the 8008 from Intel, this 40-pin "superintegrated" circuit was like no other before—it was programmable. The microcomputer age had dawned. Electronics would be changed forever.

Today, we're all computer rich, though we probably don't know it. If you can pay two dollars for a watch, you can afford to own a computer (Figure 1.4). Those "throw away" timepieces from East Asia contain a tiny microcomputer no bigger than a dime, and costing about as much to produce. If you recently purchased a car, you own 30 to 40 more computers. The auto industry is the largest user of microcomputers in the world; the average automobile contains 30 separate embedded microcomputers.

Do you have a new TV, stereo, VCR, microwave oven, washing machine, or digital clock? Add at least three or four chips for each device. As we exit one century to begin another, the average North American encounters 300 microcomputer chips in a day: in the home, office, or car. We are awash in microcomputer technology.

If you find the past exhilarating, expect even more excitement in the years to come. We'll turn to the future later in the chapter. But first, let's

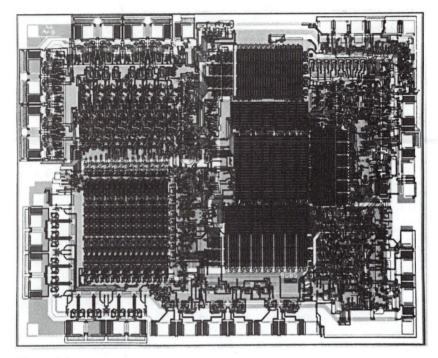

Figure 1.4
A microprocessor.
*Source: C*ourtesy of Intel Corporation.

pause to gain a better understanding of how the electronics industry is organized and classified.

Organization and Classification

Making organizational sense of any enterprise as vast as the electronics industry is quite a challenge. On what basis do we subdivide or classify all that's going on? One approach is to look closely at how the industry itself groups its activities and personnel. Another avenue is to examine how colleges and technical institutions organize their curricula in electronics and computer science. When we combine both methods, we arrive at the ten subfields shown in Figure 1.5. Although one could argue with this ten-part scheme, perhaps consolidating some subfields and separating out others not shown, the ten divisions do offer a workable framework by which to analyze the industry as a whole.

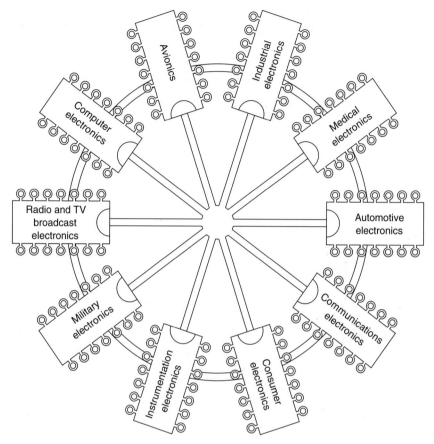

Figure 1.5
Subfields in electronics.

Before we explore each subfield, it is best to keep in mind two points. First, within each field, the design, manufacture, sales, and service of the product lines often take place on an international basis. For example, a television, within the consumer electronics subfield, may be designed in the United States, manufactured in Korea, and sold and serviced almost anywhere in the world. It is a rare electronic device today that has its origin, development, production, and use in only one country, even a country as large and wealthy as the United States.

Second, the ten subfields are intertwined. For instance, consumer electronics does not exist in a vacuum. Equipment developed may contain a microprocessor from computer electronics, a modem from communication electronics, and a sensor from automotive electronics. Furthermore, many basic technologies penetrate a number of subfields. Take lasers, for example. Do they belong in the medical, communications, consumer, computer, industrial, or military electronics fields? Or do they belong in all these categories? As you can see, the borders between subfields are indistinct, and no universal agreement exists as to exactly what should be included in each one.

That's worth keeping in mind when you settle on a career choice. Chances are, no matter what area you work in, you'll need to know something about related subfields.

Let's turn our attention now to each subfield, seeing just what it is, what it encompasses, and what type of individuals might be attracted to it as a career.

Medical Electronics

Medical electronics is concerned with electronic equipment used in the research, diagnosis, and treatment of diseases. In research, an electron microscope might be used to magnify a cancer cell 350,000 times. An electrocardiograph (ECG) is a common diagnostic tool that can detect abnormal heart rhythms. An endoscope can aid in performing microsurgery on an inner part of the body. Those working in the field of medical electronics should have some knowledge of biology as well as a strong background in electronic instrumentation.

Automotive Electronics

Automotive electronics concerns itself with how electronic sensors, actuators, and circuits are used to control automotive functions in fuel and emission control, speed control, passenger safety, and the like. In fuel-control systems, sensors measure exhaust gas, oxygen, absolute manifold pressure, etc., to obtain the optimum fuel/air mixture. Cruise-control systems, consisting mostly of analog circuitry, maintain the constant speed (velocity) of a vehicle. *Virtual* speedometer readouts, which project the digital display above the hood at eye level, enhance traffic safety by allowing the driver to maintain eye contact with the road. Individuals interested in automotive electronics should also have a liking for auto mechanics.

Industrial Electronics

Industrial electronics is a broad field, encompassing electronic equipment found on the factory floor, the engineering design laboratory, and the front office. Programmable controllers and robots are used in process control, from baking cakes to assembling automobiles. Computer-aided-design (CAD) workstations have all but eliminated the drafting table. In the office, devices from copiers to fax machines and desktop computers are revolutionizing the way information is stored, retrieved, displayed, and analyzed. Tying all this together is the concept of computer-integrated manufacturing (CIM). The industrial production facility of today is electronic- and computer-based. Anyone interested in manufacturing will require a strong background in electronics and computers.

Radio and Television Broadcast Electronics

The subfield of radio and television broadcast electronics deals with the installation, repair, maintenance, and operation of electronic equipment in live commercial radio and television broadcast stations, both in the studio and at the transmitter site. In the studio, various audio mixers and modulating equipment are installed, repaired, and adjusted. At the transmitter, high-power oscillators and broadcast antennas are maintained. In both locations, a wide range of sophisticated equipment is needed to place and keep an audio or video signal on the air. While an FCC (Federal Communications Commission) license is no longer required by law to work on such equipment, many large broadcast stations will not hire you unless you have one. If you intend to go into this important subfield of electronics, plan to obtain your FCC license as soon as possible.

Military Electronics

Military electronics includes electronic equipment for smart weapons, shipboard communications, satellite surveillance, and the like. During the Persian Gulf War of 1991, Air Force smart bombs and laser-guided missiles received much credit for the operation's success. Offshore, navy ships, equipped with the most sophisticated communications gear in the world, patrolled the gulf waters. Overhead, satellites costing as much as $500 million apiece mapped the terrain below, surveyed suspected troop concentrations, and relayed vital military data back to earth. And on land, sophisticated night-vision equipment gave the U.S. forces what one commander declared "the greatest single advantage over the enemy." Electronics played a key role in winning the war. Obviously, those interested in military electronics will get their greatest exposure by joining one of the military branches.

Communications Electronics

Communications is a very large subfield that includes audio, video, and data telecommunications (Figure 1.6). Audio involves the standard, wire-based telephone system, as well as the traditional aspects of radio: amateur, commercial, citizen band, and marine. It also embraces new modes of wireless communication, such as pagers and cordless and cellular telephones. Video, of course, means television, both broadcast and cable. Data communications is concerned with facsimile terminals, modems, network controllers, and related computer-based equipment. Tying much

Figure 1.6
Communications electronics.
Source: Cleveland Institute of Electronics.

of the communications network together are fiber-optic and satellite communication links. As the United States spends billions transitioning from analog to digital television, opportunities in communications electronics will continue to grow.

Consumer Electronics

Consumer electronics is probably the broadest subfield, with peripheral products such as burglar alarms, personal computers, and health aids spilling over into allied subfields (Figure 1.7). At its core are personal-use items for entertainment, safety, and information: television, videocassette recorders (VCRs), CD players, camcorders, home satellite receivers, calculators, cameras, telephone-answering equipment, Fax machines, microwave ovens, musical instruments, smoke detectors, and, of course,

Figure 1.7
Consumer electronics.
Source: Cleveland Institute of Electronics.

stereo equipment. Those working in consumer electronics must have good technical skills and be able to interact with customers and clients in a friendly and professional manner.

Instrumentation Electronics

To diagnose and repair today's complex electronic devices, sophisticated test equipment is often required. At the low end are the analog and digital multimeters, function generators, audio oscillators, and component checkers. The upper end is populated with logic analyzers, spectrum analyzers, microprocessor-development systems, storage oscilloscopes, and virtual instruments (computers as test equipment). Also within this subfield we increasingly encounter automated test systems, of which the in-circuit IC tester is an example. Those wishing to work in instrumentation electronics should have substantial hardware experience and be familiar with computer programming at the assembly language level.

Avionics

Avionics is an acronym designating the field of *avi*ation electr*onics*. It is the branch of electronics concerned with aviation applications. As such, it incorporates technology from related subfields: instrumentation, computer, communications, military, and even industrial electronics. Obviously, if you are interested in both aircraft and electronics, avionics is a career choice worth investigating.

Computer Electronics

Within the explosive subfield of computer electronics are products so small they are actually imbedded in other electronic equipment, such as microprocessor controllers. Or they are so large they fill entire rooms—the mainframe computer, for instance. In between you will find palm-sized, laptop, portable, desktop, workstation, and minicomputers. If you work in this area of electronics, you will also be concerned with local-area networks (LANs), wide-area networks (WANs), and various other communication links that keep these systems talking to each other. A knowledge of both hardware and software (basic computer programming) is a must for anyone seeking a career in computer electronics, be it in design, repair, installation, or maintenance.

All the previously mentioned subfields are here to stay and will experience change and growth in the years to come. To be sure, some—such as medical electronics, automotive electronics, communications electronics, and even consumer electronics—will probably grow the fastest, whereas military electronics will rise only moderately. Nonetheless, all areas will offer opportunities for those looking for rewarding careers.

Crossing into the New Millennium

So far, we have explored the electronics industry: examining its diversity and success, highlighting its twentieth century history, and identifying and defining its many subfields. In this concluding section, let's play futurist in hopes of seeing what new electronic products and services the next few years will bring.

What will our lives be like as the first decade of the twenty-first century unfolds? What electronic devices, many of which will be designed, built, tested, installed, and sold by you, will be commonplace in the time it takes for some to graduate? To be sure, if the last couple of decades are any indication, many electronic products and services we find familiar at home, work, or play in the next decade are yet to be thought of. If you doubt this, just examine what happened during the 1990s, from 1991 to 2000. Ask yourself what electronic devices existed in 2000 that were unheard of in 1991—or at least were unavailable as viable consumer products. The Internet, wireless communication devices, and digital TV are examples that come to mind, though there are many more.

In the early 1990s there were 60 million microcomputers in the United States; in 1995 there were 100 million. Today, it's close to 180 million. The microcomputer industry is a $150 billion industry, making up 2.5 percent of the gross domestic product. One-hundred thousand PCs are manufactured every day, from Malaysia to Mississippi, for world consumption. According to Brian C. Fenton, former editor of *Electronics Now,* "Personal computers have caught on faster than just about any other modern invention, and they have altered the way we work (and play) so drastically that, short of catastrophe, it's inconceivable that we might ever return to the way things used to be done."

To understand where computers are headed, we can count transistors. Moore's Law, pronounced in 1965, says that the number of transistors on a chip, and the resulting performance, doubles every 18 to 24 months. When Intel introduced the 4-bit, 4004 microprocessor in 1971, it contained 2,300 transistors. Today's Pentium chips have close to 10 million.

The 4004 ran at 2 MHz, the latest version of Pentium (as of this writing) is cruising along at 1.5 GHz. Although PCs are a product of the 1980s and 1990s, computers, as we saw earlier, actually go back to the early 1940s. Modern **facsimile machines,** however, have a more recent genesis, the first one being introduced by Xerox Corp. in 1961. But it took a convergence of many forces in the 1980s to make the fax the ubiquitous office (and, more and more, home) product of today. According to the futurist Alvin Toffler of *Future Shock* fame (from Alvin Toffler, *Power Shift*):

> Suddenly, in the late 1980s, several things came together. Fax machines could be produced at low costs. Telecommunications technologies vastly improved. AT&T was broken up, helping to cut the relative cost of long-distance services in the United States. Meanwhile, postal services decayed (slowing transaction times at a moment when the economy was accelerating). In addition, the acceleration effect raised the economic value of each second potentially saved by a fax machine. Together these converging factors opened a market that then expanded with explosive speed.
>
> By the spring of 1988, as though overnight, Americans received a hailstorm of phone calls from friends and business associates pleading with them to install a fax. Within a few months, millions of fax machines were buzzing and bleeping all over America.

Third on our short list is the **cellular phone.** The basic concept has been around since the 1940s. By the 1970s computer and control technologies had developed enough to consider a practical network. Then, according to Stephen J. Bigelow ("All about Cellular Telephones," *Radio Electronics*), the airways exploded:

> By the end of the 1970s, a prototype cellular network was implemented in Chicago. It was a hit. AT&T assumed a leading role to develop the new cellular technology. The first full-fledged cellular network went on-line in Washington, D.C. in 1984.
>
> At this time [1995], cellular services are available in every urban center and many suburban areas across the United States. It is only a matter of time until cellular service is available in all parts of the country.

Speaking of the near future, let's now turn our attention to a few easily discernible, up-and-coming electronic products and services likely to alter our lives significantly in the next decade. As these and many yet-to-be-thought-of electronic devices are developed and come to market, economic growth, new activity, new jobs, and new opportunities will open up to those choosing to be a part of the electronics industry of the future.

The **Smart Building,** be it a home, office, factory, or warehouse, will be a reality in a few years. A prototype *Smart House,* created by the NAHB Research Foundation (a subsidiary of the National Association of Home Builders) already exists. It contains a *closed-loop power system,* a *high-speed data-transmission system,* and a *logical central control system* that make the house and its appliances more user-friendly and efficient. In this Smart House, the danger of electrocution is eliminated, as are house fires caused by electrical faults and gas leaks. Appliances are monitored and controlled from any point in the house. For example, using the new communications link, if the doorbell rings while you're vacuuming the rugs, the vacuum cleaner will turn itself off so that you can hear.

Most significantly, though, dc as well as ac will be available in the Smart House of the future. As a result, according to David J. MacFadyen, president of the NAHB Research Foundation, Inc. ("The Home of the Future," *Radio Electronics*):

> Virtually every product that attaches to the various wiring systems of a home today will be changed to take advantage of the *Smart House* technology, which will offer a vastly improved array of home products and a better way of living for all of us.

In other words, huge new markets will exist for entirely new electronic products and services that run on dc.

Within the Smart House, and to some extent dependent on it, will be an entirely new consumer product called the **multimedia appliance,** an information and entertainment center for the whole family. Consisting of a powerful joystick-, pen-, and voice-operated personal computer, a digital television, and Internet downloaded entertainment and information software, this sophisticated all-purpose consumer product will make the old color TV obsolete. George Gilder had it right in his *Life After Television:*

> You could take a fully interactive course in physics with the world's most exciting professors, who respond to your questions and let you learn at your own speed. You could view the Super Bowl from any point in the stadium, or soar above the basket with Michael Jordan— all on a high-resolution display.

In the next decade, consumer electronics will have taken on a whole new meaning.

Moving from the home to the automobile, **stolen vehicle recovery systems (SVRS)** are "hot," if you'll excuse the pun. Are such systems really needed? Auto theft is perhaps the fastest growing crime in America. More than 2.2 million cars were stolen in 2000—one every 17 seconds. That comes to an $18 billion annual loss—and the rate of theft is increasing.

There are a number of SVRSs on the market, typical of which is the Teletrac spread-spectrum RF telemetry system. It can pinpoint the location of a stolen vehicle down to 100 feet. The $400 Teletrac system is triggered when the car is started without the owner's ignition key. Police then track down the vehicle and its thief, usually within minutes of being silently alerted.

But today's stolen vehicle recovery systems do far more than just protect your car—they protect you as well. In the case of a roadside emergency, you just press a button on a small keyboard, and a message is sent to your auto club telling it exactly where you are and what's wrong with the car (flat tire, out of gas, dead battery, etc.). Help is dispatched immediately. Needless to say, such systems more than pay for themselves in short order. Eventually every new car sold in America will probably be equipped with some type of SVRS. Thus we see another minielectronics industry in the making!

Then there are **Web Phones,** cellular phones that give you quick, though scaled down, access to the World Wide Web. Industry experts estimate that by the end of 2001, 100 million phones with Web access and email features will have been sold worldwide. That, perhaps surprisingly, is about 12 times the number of hand-held computers. In Japan, at the end of 2000, cell phones had already become the primary connection to the Internet. By 2005, 48.6 milliion U.S. households are expected to have Internet cell phones. And by 2004, 1.3 billion people worldwide will have such devices.

True, we are not talking about video streaming and the like. These Web-connect devices are for information that is personalized, and fast. What the user wants to know is: How much? Where? What flight? When? Can I book it?

Still, as one commentator put it, "The wireless Internet in the U.S. is the second coming of the Internet." Interestingly, in some parts of the world, it's the first coming.

Finally, it is, perhaps, in the arena of *medical electronics* that we may well see the most dramatic and beneficial advances of the next decade. And, given the ever-increasing mobility of the world's population, storage and retrieval of an individual's medical records will be vital. Here's what Dr. Ray Fish, Ph.D., M.D., sees in this regard ("Medical Technology in the 21st Century," *Radio Electronics*):

Some people will carry copies of their medical records on wallet-size digital discs having a standard format readable by any hospital computer. Others will have their entire medical history stored on a programmable pinhead-size optical memory chip that is concealed under a dental filling. In the event of an accident, a laser-scanner aimed at a bicuspid, an incisor, or even dentures will spew out the patient's medical history.

Statements such as this make it clear that the advanced medical practices in the next few years will be due mainly to advances in electronics.

Of course, these new products and services will just be the beginning. Space doesn't permit us to detail others, such as penny-sized nano-CDs with 800 times the capacity of a standard CD, handheld scanners that can see through walls, home-based handheld biological sensors that can detect bacteria-contaminated food before a family eats it, cellular videophones, and widespread installation of on-board navigation systems for your car, to name a few.

In this last section, and indeed throughout the chapter, we have presented an upbeat and optimistic view of the electronics industry: where it has been, its economic status today, and its outlook for the future. We do so because we believe in its strength and its promise.

"Worker-Starved Companies Go Hire and Hire," blares one recent news headline. "U.S. Jobless Rate Dips to 4.3%, a 28-Year Low," declares another. And "High-Tech College Students in High Demand," trumpets a third. As the twenty-first century unfolds, being part of high-tech is in.

For example, at a recent recruiting conference, Northwestern University placement director William Banis casually mentioned a college freshman he had hired to maintain his office computer network. He was astonished when dozens of companies asked for the student's name.

In early 2001, young workers with prized, high-tech job skills enjoyed the hottest hiring market in 32 years. Employers expected a 24 percent increase in openings.

Those statistics are a far cry from the gloomy picture just a decade ago. In an earlier edition of this text, in this very space, we were forced to announce corporate downsizing and employee layoffs. Yes, it could happen again. You can't have economic stability and economic growth at the same time. And economic growth means economic change. With the latter there will always be winners and losers. Some companies and individuals will gain and find opportunity; others will lose and fall by the wayside. The winners need to be highly trained and educated, be flexible and adaptable, and, in some cases, be willing to accept assignments anywhere in the world. Exactly who will populate the offices, laboratories, and factories of the electronics industries in the years to come? In Chapter 2 we'll find out.

Summary

In Chapter 1 we explored the omnipresent electronics industry, examining its diversity and success. We looked at its history, from the vacuum tube to the Internet. And we saw how electronics is organized and classified. Finally, we peeked into the future, with an eye for the electronic products and services that may spring forth as you prepare for a career in electronics.

Review Questions

1. Electronic products contain electronic _____, connected to form electronic _____, that in turn control the flow of _____, or electricity.

2. Today, electronics is the world's _____ industry.

3. Today, according to one government report, the high-tech industry accounts for more than _____ percent of the national output of goods and services.

4. In January, 1998, virtually every U.S. household had a

   _____ _____.

5. _____ is an ambitious 288-satellite project that will revolutionize the telecommunications industry.

6. In 1947, the vacuum tube met its eventual replacement, the

   _____.

7. _____ electronics is a broad field, encompassing electronic equipment found on the factory floor, the engineering design laboratory, and the front office.

8. _____ is an acronym designating the field of *avi*ation and electr*onics*.

9. Today, close to _____ million microcomputers are in use throughout the United States.

10. Those working in the electronics industry need to be highly _____ and _____, be flexible and adaptable, and, in some cases, be willing to accept assignments anywhere in the world.

Individual and Group Activities

Summarize the results of each activity in a 2- or 3-page report or a 5-minute oral presentation to the class.

1. Pick one of the ten electronics subfields discussed in this chapter. Find out more about it, with particular emphasis on job growth and career opportunities.

2. Pick four or five of the electronics subfields discussed in this chapter and survey their growth potential in your geographical area. Make use of the telephone or in-person visits to gather information.

3. Investigate the history of a major electronic product or service, such as radio, television, the computer, fax machine, or compact disc. Emphasize the economic growth of the product or service since its commercial introduction.

4. Investigate the history of a major electronic innovation, such as the vacuum tube, transistor, integrated circuit, microprocessor, or laser. Who were the people behind the development? How was the development initially received by the public? How long did it take to go from invention to practical application?

5. Visit a museum of science, industry, and technology. What electronics innovation attracted your attention? Describe what it is, its influence on our lives, and the career opportunities open to those who designed, developed, manufactured, tested, repaired, and sold the product or service.

6. Contact a local radio or television broadcast station and ask for a tour of the technical facilities. Pay particular attention to the types of equipment used and the job training required to operate and maintain the equipment. Note the working conditions encountered by the technical staff.

7. Spend a day with an auto mechanic. Determine what portion of his or her work involves electronics. Is the mechanic just replacing electronic modules, or is a broad knowledge of electronics required?

Issues for Class Discussion

1. Discuss an electronic product or service that didn't become popular, such as the picture phone (as of yet) or self-guided automobile, and see if you can determine why. Was it a technological failure or socioeconomic problem?

2. Discuss an electronic product or service you think is destined for success in the next few years. What factors do you think will contribute to its acceptance? What job opportunities will be generated as a result of this innovation?

3. Discuss what is meant by the statement: "You can't have economic stability and economic growth at the same time." Do you believe this is true? If so, what implications might there be for the electronics industry? For your future?

CHAPTER

2

Electronics:
Field of Dreams

Objectives

In this chapter you will learn:

- About those employed in the electronics workplace.
- What kind of careers are available for electrical engineers.
- How the career of electronics technician differs from that of electrical engineer.
- About the salaries for electronics technicians, electronics technologists, and electrical engineers.
- The job prospects for those working in electronics.
- About the workplace revolution.
- How you can be part of the symbolic-analytic services career category.

In the 1968 motion picture *The Graduate,* Dustin Hoffman, the young, somewhat bemused, career-searching college graduate, is cornered at a cocktail party by a well-heeled upper-management type. In the most famous line of the film, the pushy manager exclaims: "Plastics my boy, plastics, that's where the future lies—in plastics!" Encouraged to shun the more traditional business career opportunities in cosmetics, automobiles, or publishing, the hero of the film is urged to lock on to the new up-and-coming plastics industry as the smart career choice for the decades to come.

If that film were produced today, however, it is a fair bet that electronics, not plastics, would be the clarion cry from the clairvoyant executive. Never before has a field so captivated the public's enthusiasm, veneration, and imagination. Never has an activity permeated our lives, be it at home, work, or play—or in the hospital or on the battlefield. And in no other occupation have career possibilities seemed so diverse, so accessible, and so challenging. Whether you're "the graduate" or not, electronics seems to stretch forth, unimpeded, into a field of dreams.

In this chapter, we look more closely at electronics as a career. We begin by exploring technical careers in the electronics workplace: the role of the engineer, technologist, technician, and assembler. Next, we see how these and related careers are likely to develop in the next few years. We pay particular attention to the employment outlook for each career category. Finally, we investigate the so-called workplace revolution, where we see what the term *global economy* means, how job opportunities in the United States are affected by it, and what job skills, aptitudes, and knowledge you will need no matter what electronics career path you choose.

The Electronics Workplace Today

In Chapter 1 we saw a vast, diverse, and growing electronics industry. The people employed by this huge enterprise can, broadly speaking, be divided into three occupational groups. There are the **technical occupations,** where the majority of one's time is spent in doing electronics, from circuit-board assembly to high-level solid-state research. There are the **nontechnical occupations that require some technical background,** such as technical sales, technical writing, technical training, and supervisory/managerial activity. And, of course, there are **nontechnical occupations** essential to support any firm. We refer here to accountants, secretaries, marketing personnel, maintenance workers, and the like. For this latter group, formal training in electronics is usually not necessary.

In this chapter, we concentrate on the first group—technical occupations. In Chapter 3, "The Electronics Technician: Bringing Electronics to Life," we explore the specific technical career of electronics technician in considerable depth. Pursuits such as accounting and finance are not the subject of our study.

As a community college instructor, I spend a great deal of time speaking at high-school career days. Invariably, a group of 15 to 50 students, professing an interest in electronics, is gathered in front of me, eager (in some cases) to hear my words. When I ask them what they want to do in electronics, they announce: "I want to be an electronics engineer." Further inquiry usually reveals that they know of no other career possibilities within electronics. For most of them, being an electronics engineer and being in electronics are the same thing.

Being an **electronics engineer,** however, is not the only choice open to you. There is the **electronics technologist,** the **electronics technician,** and even the **electronics assembler.** Presented in that order, from engineer to assembler, we find ourselves moving from the theoretical to the practical, from an occupation requiring at least 4 years of college to one demanding no more than a high-school diploma or certificate. Let's look at the four options in turn.

Within the broad field of engineering, three subfields concern themselves with electricity. These are **electrical, electronic,** and **computer engineering.** Individuals declaring themselves to be one of the three will usually have an appropriate bachelor of science degree in electrical engineering. (Electrical engineering, first recognized as a career in 1884, is the historical term for what is now called electrical, electronic, and computer engineering.) Responsibilities include specification, design, development, and implementation of products or systems, as well as research to create new ideas. These positions stress theory, analysis, and design. Any one of the subfields can lead to allied fields such as biomedical engineering and aerospace engineering.

But what, you may ask, is the difference between the three subfields? Aren't electrical and electronic engineering the same thing?

Although all three share the commonality of working with electricity, there are clear differences in emphasis. According to Dr. Rudolf A. Stampfl, Staff Director, Educational Activities Department of the Institute of Electrical and Electronics Engineers (IEEE), in a letter to the author:

> An electrical engineer is expected to know electrical power generation and distribution engineering. This means he [or she] is knowledgeable of generators, motors, transformers, transmission lines, and high-voltage technology. An electronics engineer is not necessarily expected to know these subjects. Additionally, an electrical engineer may know the application of some electronics instruments and controls but may not be interested in their design.

An electronics engineer should know of circuit design, transistor circuits, logic design, communication engineering (modulation source coding), microwave circuitry, e.m. wave propagation, antennas, control systems, etc.

The computer engineer, on the other hand, has knowledge and experience not only in electricity and electronics, but in computer architecture, switching theory, computer design, numerical methods, database design, operating systems, artificial intelligence, voice communications, and the like.

Typical job titles for all three subfields are as follows:

Design engineer

Project engineer

Engineering specialist

Network designer

Chief engineer

Quality-control engineer

Software engineer

Development engineer

Reliability engineer

Research engineer

Systems design engineer

Field engineer

Test engineer

Network administrator

Sales engineer

Moving along the continuum from theory to practice, we next encounter the **electronics technologist.** He or she also has a bachelor's degree (in some cases, an associate degree in engineering technology), such as a bachelor of engineering technology (BET), bachelor of science in engineering technology (BSET), or a bachelor of science in applied technology. The technologist's education and occupational duties involve less theory and more practical orientation than that of the electrical engineer. The technologist is more concerned with applications—the how as opposed to the why. The engineer uses theory and design methods to develop products and systems. The technologist takes a concept and transforms it into a prototype or product. Engineering technology stresses the application of today's technological know-how to current industrial practices and design procedures. The occupation of electronics technologist, or ET, is well established in Europe and Japan. It has come into its own in this country only in the past three decades. Typical job titles for electronics technologists are as follows:

Electronics technologist

Biomedical engineering technologist

Sales engineering technologist

Customer service engineering technologist

Service engineering technologist

Systems test engineering technologist

Product engineering technologist

Software engineering technologist

Documentation engineering technologist

Quality-control engineering technologist

Applications engineering technologist

R&D technologist

Engineering assistant

Internet administrator

Next, we come to the technical occupation of **electronics technician.** Since this career category is the focus of our book, we will devote an entire chapter to a discussion of its characteristics (Chapter 3). For now, we can say that an electronics technician is generally required to complete one to two years of specialized education, usually leading to an associate degree. Electronics technicians are the hands-on part of the design and manufacturing team. They typically install, test, and maintain products in the field or in the plant (Figure 2.1). An electronics technician is often called upon to confirm specifications or operations as originally designed. Such technicians do breadboarding, model building, programming, testing, troubleshooting, system configuration PC-board design and layout, etc. Typical job titles for the electronics technician are as follows:

Service technician

Manufacturing specialist

Field service technician

Figure 2.1
Electronics technician.
Source: Cleveland Institute of Electronics.

Customer service representative

Test technician

Bench technician

Calibration/lab technician

Network installer

Electronics assembly, while certainly part of the electronic product/system picture, does not, strictly speaking, represent a career path, but rather a job. The difference between a career path and a job is that the latter belongs to your company, but the former belongs to you.

Electrical and electronic assemblers wire and mount components, combine them into subunits, and fit them into housings that make up a complete product. They use light hand or power tools, such as wire strippers and crimpers, soldering and microwelding tools, and small wrenches. Knowledge of electronic circuit design and operation is not required.

While jobs for electronic assemblers are diminishing, working as an assembler is an excellent way to gain exposure to the electronics industry. A part-time job as an assembler while attending school could be invaluable.

You want answers to six key questions concerning any occupation. You want to know (1) the job description, (2) the working environment, (3) earnings and benefits, (4) additional career opportunities and promotions to which the job might lead, (5) the education and training required, and (6) the employment outlook, near and long term. Table 2.1 lists the four technical occupations discussed earlier—electrical engineer, electronics technologist, electronics technician, and electronics assembler—and matches them against five of the six job issues just mentioned. The sixth topic, employment outlook, is discussed in the next section. Before we do that, however, let's elaborate a bit on the information presented in Table 2.1.

As we have seen, electrical engineers are the designers on the technical team. They do the research, make the schematic sketches, and, in many cases, perform the experiments necessary to determine if a product or service idea is feasible. They work in laboratories and technical facilities as leaders and planners for the technical staff. Their earnings vary widely, depending on experience, responsibilities, the size of the company, and location. Salaries presented in Table 2.1 are averages for the entire country and are for engineers whose work is primarily technical. Those who have moved into management, sales, or high-level research or who accept assignments overseas or in hardship areas will usually earn more. And, of course, engineers with advanced degrees, masters or Ph.D.s, may command much higher salaries. In sum, electrical engineers (all three types), highly educated and experienced, are well-paid professionals working in technically challenging and pleasant environments. (Though earnings in all categories are for the year 1998, you can safely add 4 percent for every year since then.)

Electronics technologists, on the other hand, have a problem—they often suffer from an identity crisis. They hear statements such as, "Are you a *real* engineer?" or, "Aren't you just an electronics technician with a bachelor's degree?" Those who do understand the job description recognize that the technologist is usually an engineer, but one with a practical bent. As engineers they are concerned with high-level issues regarding technology already in existence. Their working environment, however, is less in the laboratory and more on the production floor. They are out there working with the materials and equipment that produce electronic devices. They may walk around with a little dirt under their fingernails.

As Table 2.1 shows, electronics technologists are often paid as much as 25 percent less than electrical engineers. On the positive side, however, this

TABLE 2.1
Technical occupations.

Career/ Job Title	Job Description	Working Environment	Earnings and Benefits	Additional Career Opportunities	Education and Training
Engineer: electrical, electronics, computer	Specification, design, development, and implementation of product or system, as well as research to create new ideas	Most work regular hours in clean, professional surroundings. Respect-ed as leader of product-development team.	Earnings (1998): • Ave $51,700 • Max. $79,250 • Min. $21,992 Benefits: Usually full company benefits, including stock options.	Management Sales Research Own company	Bachelor of science in electrical engineering. Advanced degrees often required for advanced work.
Electronics technologist	Applications of theory, analysis, and design. Applies today's technolo-gical know-how to current industrial practices and design procedures.	Most work regular hours either in a laboratory or on production floor in a hands-on environment. Sometimes confusion exists as to actual role.	Earnings (1998): • Ave. $38,137 • Max. $57,645 • Min. $21,323 Benefits: Medical, dental paid vacation.	Field engineer Marketing Sales Process management Quality assurance Customer service Supervision Documentation Training	Bachelor of engineer-ing technology (BET) or bachelor of science in engineering technology (BSET).
Electronics technician	Fabrication, operation, testing and trouble-shooting, and maintaining existing equipment. Building prototype products and systems.	Most work regular hours in labs, offices, elec-tronics shops, industrial plants or construction sites. Service techni-cians usually spend time working in customer establishments.	Earnings (1998): • Ave. $33,743 • Max. $51,000 • Min. $28,000 Benefits: Medical, dental paid vacation.	Technical sales Technical writing Technical training Supervisor manager Customer service	Minimum: 1-year certificate at trade school or community college. Usually associate degree is required.
Electronics assembler	Wire and mount compo-nents, then fit them into housings which make a complete product.	Most work in electronics assembly plants. Jobs are routine and repetitive. Little opportunity to display creativity and initiative.	Earnings (1998): • Ave. $10.66/hour • Max. $19.10/hour • Min. $5.58/hour Benefits: In most cases, medical, dental, paid vacation.	With additional schooling can advance to entry-level electronics technician position.	Minimum skills: Ability to use handtools and solder. Certificate at occupational skills center, GED, or high-school diploma desirable.

Source: U.S. Department of Labor

group tends to have greater additional career opportunities than those of electrical engineers. Given their practical and people-skills orientations, technologists are in an ideal position to move into supervisory/managerial roles or to take on responsibilities in customer service, in-house or customer training, or sales and marketing. Having a college degree in a technical field, electronics technologists can move in many career directions.

Electronics technicians help develop, manufacture, and service electrical and electronic equipment such as two-way radios, telephones, fax machines, industrial and medical measuring or control devices, navigation equipment, computers, and computer networks. They are the true hands-on types—they work directly with electronic components, circuits, and systems.

Some electronics technicians are out in the field, constantly on the move from one customer location to the next. These field techs are often under considerable pressure to get faulty equipment up and running as quickly as possible. Frequently surrounded by harried, impatient office or factory personnel, their people skills are repeatedly put to the test.

In-house (or bench) techs work in a more controlled, usually less pressured, environment. However, they are frequently called upon to tackle more complex technical problems. These people are often the "supertechs," who are valued primarily for their extensive troubleshooting skills.

Of the four technical fields discussed here, the pay scale for electronics technicians probably varies the most. In 1998 the average income for an electronics technician was $33,743. But the "supertech" had an average income of no less than $51,000. And 10 percent of all technicians made more than $50,000. We'll look more closely at the issue of salary and benefits for electronics technicians in the next chapter.

As with the electronics technologist, being an electronics technician can act as a springboard to other careers. Technical sales, training, and writing have already been mentioned. And for those techs with an associate degree and an IT certification, such as A+ certification, there are further opportunities as well.

Electronics assemblers, as the title indicates, assemble electronic devices. Automation and off-shore labor have reduced the demand for such workers here in the United States. For example, Connie Chavez, of CLS Industries, knows first-hand what automation means: "I've worked here at the laser company for the past 5 years. Six months ago, 21 women were stuffing and soldering circuit boards. See that flow solder machine in the other room? It was installed in March, 3 months ago, and now there are only 15 women left. It's because of that machine, I know it is." Nonetheless, those who are working do so in a reasonably comfortable factory environment, and, in 1998, were paid an average of $10.66 an hour. Without additional training or education, however, advancement is limited.

Electronics Careers in the New Millennium

It is now time to peer ahead, to examine the near-term employment outlook for electronics careers. In doing so, we'll look first at demographics, that is, the characteristics of the unfolding national workforce. Next, we'll investigate the goods-producing versus service-producing industries, see how the balance between the two is shifting, and check the effects of this shift on the job outlook for technical careers. We will then turn specifically to job projections for the four electronics careers discussed in the previous section. We'll pay particular attention to what's likely to unfold for the electronics technician in the coming few years.

The U.S. population grew more slowly during the late 1990s than it did during the 1980s. The same is true for the labor force, which comprised people who were either working or looking for work. The civilian labor force is expected to increase by 17 million, or 12 percent, to 154.6 million over the 1998–2008 period. This increase is almost the same as the 13 percent increase during the 1988–1998 period but much less than the 19 percent increase during the 1978–1988 period.

The gender, racial, and ethnic characteristics of the work force are also changing. Women were only 41 percent of the labor force as recently as 1976; by the year 2008 they are expected to account for 47.5 percent. White non-Hispanic men, on the other hand, are a shrinking part of the U.S. labor market. By 2008, blacks, Hispanics, Asians, other nonwhite racial groups, and women will account for the majority of labor-force entrants.

The youth labor force, ages 16 to 24, is expected to slightly increase its share of the labor force to 16 percent in 2008, growing more rapidly than the overall labor force for the first time in 25 years, according to the U.S. Labor Department. The large group of workers 25 to 44 years old, who comprised 51 percent of the labor force in 1998, is projected to decline to 44 percent of the labor force by 2008. Workers 45 and older, on the other hand, are projected to increase from 33 to 40 percent of the labor force between 1998 and 2008, due to the aging baby-boom generation.

What, then, do these demographics mean for the technical job outlook in the years to come? According to the U.S. Department of Labor *(Occupational Outlook Handbook 2000–2001),* "Employment of technicians and related support occupations is projected to grow by 22 percent, adding 1.1 million jobs by 2008."

Also contributing to a brighter employment outlook, for the technically gifted, is the need for personnel replacements. Again, the U.S. Department of Labor:

> Job openings stem from both employment growth and replacement needs. Replacement needs arise as workers leave occupations. Some transfer to other occupations while others retire, return to school, or quit to assume household responsibilities. Replacement needs are projected to account for 63 percent of the approximately 55 million job openings between 1998 and 2008. Thus, even occupations with slower than average growth or little or no change in employment may still offer many job openings.

And then, of course, there are the new technical occupations ready to open up. According to Clyde Helms of *Changing Times,* "There will be plenty of jobs in the years ahead in entirely new occupations built around such technologies as communications, computers, robotics, biotechnology, electric power generation and storage, fuels, and materials. . . . We think one of the major jobs in the future will be the robot technician."

Beyond demographics, the mix between goods-producing and service-producing industry jobs will have an important impact on the job picture in the next few years. It's fashionable to see manufacturing in this country as a dying enterprise, with closed rust-belt plants and jobs fleeing overseas to workers ready to work for a pittance. But that picture can be deceptive. As a percentage of our gross national product (GNP), manufacturing has held steady at 41 percent since World War II. And as a result of new plants and equipment introduced during the industry restruction of the 1990s, productivity in the manufacturing sector has increased 5 percent a year since 1990. Now U.S. manufacturers are adding jobs and exporting to the global market. The U.S. Department of Labor states *(Occupational Outlook Handbook 2000–2001):*

> Manufacturing employment is expected to decline by less than 1 percent from the 1998 level of 18.8 million. The projected loss of jobs reflects improved production methods, advances in technology, and increased trade.

Meanwhile, in the service sector, jobs are expanding even faster: 34.5 million were employed in 1988, 46.3 million were working by the year 2000. But aren't jobs found here just for sales clerks, "hamburger flippers," or lawyers? Hardly! Again, we quote the U.S. Department of Labor *(Occupational Outlook Handbook 2000–2001):*

> Although service-sector growth will generate millions of clerical, sales, and service jobs, it will also create jobs for financial managers, engineers, nurses, electrical and electronics technicians, and many other managerial, professional, and technical workers.

Technical people need not fear a shift from goods-producing to service-producing industries. On the contrary, they should welcome it.

In addition, computer-related jobs are expected to grow the fastest over the period between 1998 and 2008. In fact, these jobs (computer engineer, computer support specialist, computer systems analysts, and database administrator) make up the four fastest growing occupations in the economy, with 108, 102, 94, and 77 percent increases, respectively.

But specifically, what's happening on the electrical engineering, technologist, technician, and assembler levels in terms of job numbers? Table 2.2, compiled from U.S. Department of Labor data, tells the story.

Employment opportunities for electrical and electronics engineers (including electronics technologists) grew at a *much faster than average rate.*

For engineering technicians, of which electronics technicians are a subgroup, growth has been *faster than average* during the same period—28 percent.

Precision assemblers, on the other hand, have seen a decrease in their ranks, going from 354,000 in 1988 to 262,000 at the turn of the century—a loss of 92,000 jobs, or a 26 percent reduction.

Overall, 18,000 new electronics jobs were added each year. For qualified electronics technicians, the numbers reached 406,000 in 1996.

There is also something else going on here that will affect technical opportunities for U.S. workers in the near future—a technical brain drain of foreign students and workers in the United States choosing to return to their countries of origin. Many schools and high-tech centers, such as Stanford University and Silicon Valley, have become global magnets for foreign professional talent. Not all this talent will stay in the United States; much of it will be lured back home. This prospect, plus increased retirement rates in the coming years, could spell trouble for U.S. industries. According to Paul Saffo, research fellow at the Institute for the Future in Menlo Park, California ("U.S. Risks High-Tech Talent Gap," *Los Angeles Times*):

> A generation of scientific and technical professionals hired in the 1950s and 1960s will retire in the next decade, causing overall retirement rates to increase 15%. Throw in declining enrollments and rising industry needs, and the long-feared professional gap could become a reality despite the contribution of foreign residents.

TABLE 2.2
Technical job numbers.

Cluster Subgroup Occupation	Estimated employment, 1988	Percent change in employment, 1988–2000	Numerical change in employment, 1988–2000	Employment prospects
Electrical and electronics engineers	439,000	40	176,000	Increased demand for computers, electronic consumer goods, communications equipment, and other electrical and electronic products is expected to result in much faster than average growth. Opportunities should be favorable.
Engineering technicians	722,000	38	203,000	Well-qualified engineering technicians should experience very good opportunities. Anticipated increases in spending on research and development and continued rapid growth in the number of technical products are expected to result in faster than average growth.
Broadcast technicians	27,000	−31	−8,500	Because of labor-saving advances, such as computer-controlled programming and remote control of transmitters, employment is expected to decline. Competition is expected to be keen in major metropolitan markets.
Precision assemblers	354,000	−26	−92,000	Despite increasing manufacturing activity, a decline in employment is projected because factories will be more automated and more products will be assembled overseas.

Source: U.S. Department of Labor.

While this could represent a problem for employers, for you as a future technical employee in the United States, it is, of course, good news. Again, more demand and reduced supply means jobs are plentiful.

The Workplace Revolution

The following are facts of interest about U.S. workers:

- Among Motorola job applicants, 20–40 percent flunk an entry-level exam requiring seventh- to ninth-grade English and fifth- to seventh-grade math.
- In an international study of 13-year-olds, the rank of U.S. students in math proficiency is last; the rank of South Korean students is first.
- Twenty percent of American workers are functionally illiterate.

This is depressing news, which points to a key element in the workplace revolution sweeping the country, indeed the world: the **knowledge gap.** This gap occurs because jobs demand more, but applicants provide less. Everywhere you turn, employers are crying the same sad tune. "Among workers, problem solving, analytical skills, and teamwork are in high demand—and short supply," say Alecia Swasy and Carol Hymowitz, writing in *The Wall Street Journal.* "Every fifth person now hired by American industry is both illiterate and innumerate," adds Dale Mann, a professor at Columbia University's Teachers College. He further declares: "Technology goes up, ability goes down." Note the singling out of technology: It is a theme echoed by everyone. According to Albert R. Karr ("Wanted: Technical Skills," *Industry Week*):

> Already, much of the job market is going higher-tech, requiring better-educated and trained workers. At the same time, the supply of workers is shifting: The job market is seeing the arrival of the "baby bust" population and an increasing number of poorly schooled job candidates. The result is a mismatch between job seekers and jobs.

In the workplace revolution of the new millennium there will be jobs that are "in" and jobs that are "out"—and the list may surprise you. Nurses, skilled craftspeople, and computer wizards are in. As Karr declares, with regard to the latter, "We need more nerds. Companies gripe about a dearth of systems designers and data processing–equipment repairers." Occupations that are out, for which demand will be down in the next decade, are production assemblers (no surprise here), doctors, and dentists. Yes, doctors and dentists will be in oversupply.

What does all this mean? It means simply that the technologically illiterate, low-skilled workers of yesterday will have no place in the work force of tomorrow or, indeed, today. The Hudson Institute report titled *Workplace 2000,* sums it up with some revealing statistics: It is also worth noting here, that, according to the *Occupational Outlook Handbook, 2000–2001,* employment in occupations requiring an associate degree is projected to increase 31 percent, faster than any other occupational group categorized by education and training.

> By the year 2000, below-average skills will be good enough for only 27% of jobs created between 1985 and then, compared with 40% of the jobs existing in the mid-1980s. And 41% of the new jobs will require average or better skill levels, up from 24%.

Let's turn from mismatch to "Matching Yourself with the World of Work," the title of a report in the *Occupational Outlook Quarterly.* Using data supplied in the *Quarterly,* we match 13 occupations in electronics with 17 occupational characteristics and requirements. The results are shown in Table 2.3. Note, in particular, the following:

1. The persistence of the *problem-solving/creativity* (3) and *initiative* (4) criteria
2. The demand for *frequent public contact* (6) and *manual dexterity* (7)
3. The lack of *confinement* (11) in the working environment
4. The variation in *earnings* (14), *employment growth* (15), and *entry requirements* (17)

Take time to study Table 2.3 closely. You might put an X by the skills you have or feel you can acquire, and then match those skills to each of the 13 occupations listed. Is a career in electronics for you?

The workplace revolution can be seen in slightly broader terms than just described by exploring the thoughts presented by Robert B. Reich, former U.S. Labor Secretary, in his lucid and penetrating book, *The Work of Nations.* Dr. Reich sees three broad categories of work emerging in the decades to come. He identifies them as *routine production services, in-person services,* and *symbolic-analytic services.*

TABLE 2.3
Matching yourself with the world of work.

Occupations	Job requirements								Work environment					Occupational characteristics			
	1	2	3	4	5	6	7	8	9	10	11	12	13	14	15	16	17
Electrical and electronics engineers		•	•	•	•									H	H	206	H
Electrical and electronics technicians			•		•		•							M	H	202	M
Commercial and electronics equipment repairers			•	•		•	•							L	M	8	M
Communications equipment mechanics			•	•		•	•							M′	L	3	M
Computer service technicians			•	•		•	•							M	H	28	M
Electronic home entertainment equipment repairers			•	•		•	•		•				•	M	M	7	M
Home appliance and power tool repairers			•	•		•	•							L	M	9	M
Line installers and cable splicers			•		•	•	•	•	•					M	M	24	L
Telephone installers and repairers			•		•	•	•	•	•					M	L	−19	L
Heating, air-conditioning, and refrigeration mechanics			•				•		•	•				M	M	29	M
Office machine servicers and repairers			•	•		•	•			•				M	M	5	M
Broadcast technicians			•		•		•	•						M	H	5	M
Electricians			•		•		•	•	•	•				H	M	88	M

1. Leadership/persuasion
2. Helping/instructing others
3. Problem-solving/creativity
4. Initiative
5. Work as part of a team
6. Frequent public contact
7. Manual dexterity
8. Physical stamina
9. Hazardous
10. Outdoors
11. Confined
12. Geographically concentrated
13. Part-time
14. Earnings
   L = lowest (10% or less)
   M = middle (11–19%)
   H = highest (20% or more)
15. Employment growth
   L = lowest
   M = middle
   H = highest
16. Number of new jobs 1984–95 (thousands)
17. Entry requirements
   L = High school or less education is sufficient.
   M = Post-high-school training, such as apprenticeship or junior college, or many years of experience to qualify fully.
   H = Four or more years of college usually required.

Routine production services entail performing repetitive tasks that are done one step at a time. Jobs in this category go beyond traditional blue-collar work. As Reich points out, "Few tasks are more tedious and repetitive, for example, than stuffing computer circuit boards or devising routine coding for computer software programs." In 1990, routine production work was performed by roughly one-fourth of the U.S. work force, and the number was declining, primarily because such jobs were being exported overseas.

In-person service, as the name implies, is work performed in direct contact with the ultimate beneficiaries of the service. It, too, can entail simple and repetitive tasks. Nurses, waitresses, janitors, cashiers, childcare workers, secretaries, auto mechanics, and security guards are just a few occupations that fall into this second category. In 1990, about 30 percent of all work done in America was in-person services, and the numbers of these workers were growing.

Symbolic-analytic services are performed by people who call themselves design engineers, software engineers, sound engineers, public relations executives, lawyers, investment bankers, and real estate developers, to name but a few. They identify and broker problems by manipulating symbols. Surprisingly, perhaps, they can also include technologists and technician types working out on the production floor. According to Reich,

> There is ample evidence, for example, that access to computerized information can enrich production jobs by enabling workers to alter the flow of materials and components in ways that generate new efficiencies. Production workers empowered by computers thus have broader responsibilities and more control over how production is organized. They cease to be "routine" workers—becoming, in effect, symbolic analysts at a level very close to the production process.

Symbolic analysts made up about 22 percent of the U.S. workforce in 2000. Though their numbers were growing at a slower rate than in previous years, they constitute the top layer in the U.S. workplace.

What does all this mean for those contemplating a career in electronics? Just this: For a career with excellent employment prospects, good salary and compensation, interesting and challenging work, and, yes, a feeling of status and self-worth, you need to look to the third employment category, *symbolic-analytic services.* Electronics engineers, electronic technologists, and even electronic technicians who have the right education, training, and experience can find themselves in this elite third group. For them, electronics is a wide-open field of dreams.

Summary

In Chapter 2 we saw how the careers of electrical engineer, electronics technologist, and electronics technician differ. We examined the salary range for each career. And we saw how those careers are changing in the new millennium. We then examined the workplace revolution, seeing how the symbolic-analytic services career category is open to you as an electronics technician.

Review Questions

1. Within the broad field of engineering, three subfields concern themselves with electricity: _____ engineer, _____ engineer, and _____ engineer.

2. The _____ education and occupational duties involve less theory and more practical orientation than that of the electrical engineer.

3. An electronics _____ is generally required to complete one or two years of specialized education.

4. In 1998 the average income for an electronics technician was _____ a year.

5. _____-related jobs are expected to grow the fastest over the 1998–2008 period.

6. Employment in occupations requiring an _____ degree is expected to grow the fastest of an occupational group categorized by education and training in the years 1998 to 2008.

7. The career of electronics technician is expected to grow _____ than the average during the next few years.

8. The _____ gap is a key element in the workplace revolution sweeping the country, indeed the world.

9. _____ - _____ _____ is work performed in direct contact with the ultimate beneficiaries of the service.

10. _____ - _____ services are performed by people who call themselves design engineers, software engineers, sound engineers, etc.

Individual or Group Activities

Summarize the results of each activity in a 2- or 3-page report or a 5-minute oral presentation to the class.

1. Interview an electrical engineer. When doing so, find out how he or she got interested in this career, what the job duties are, what the strong and weak points of being an electronics engineer are, what a typical day is like, and, from the engineer's perspective, what the job outlook in the near and long term is.

2. Interview an electronics technologist. Ask the same questions as in activity 1, but, in addition, find out how his or her job differs from that of an electrical engineer.

3. Interview an electronics technician. Ask the same questions as in activity 1, but, in addition, probe the relationship of the technician with the engineer he or she may be working with. Is the technician planning on becoming an engineer?

4. Survey a medium-size or large electronics company to find out what mix of electronics engineers, technologists, and technicians it employs. Find out how these three types of workers interact on the job.

Issues for Class Discussion

1. Since everyone agrees that production jobs in the United States are disappearing, what type of technical work, if any, will be available in the manufacturing sector in the foreseeable future? Is manufacturing important to the economy of this country? Should America even try to gain back its competitive edge in the manufacturing sector?

2. How will rising demand and reduced supply of technical workers affect your job prospects after graduation? Does it mean it will be easier to find a job even if your qualifications are less than those demanded by industry? Or will industry simply import foreign workers or ship the entire facility overseas to get the job done?

3. Using the information in Table 2.3, discuss the job requirements, work environment, and occupational characteristics for the electronics careers listed. Do you agree or disagree with the criteria presented for each occupation? How would you change the criteria?

4. In industry, you will often find engineers classified as technicians and technicians classified as engineers. For example, at CBS in Los Angeles, all technical personnel, regardless of their educational background, are called technicians. Why do you suppose this cross-defining takes place? What does it mean? How important is an occupational title anyway?

5. If electronic assembly is a shrinking occupation, would it still be of any value for future electrical engineers, electronic technologists, and electronic technicians to do any electrical assembly of their own, such as building electronic kits? What can one gain from such activity?

6. There are those who say the education of an electrical engineer is too theoretical. There are engineers who graduate not knowing how to use an oscilloscope or soldering iron. Do you know this to be true? If so, why do you suppose this is the case? What, if anything, should be done about it?

CHAPTER 3

The Electronics Technician:
Bringing Electronics to Life

Objectives

In this chapter you will learn:

- About the demand for electronics technicians.
- How electronics technicians are classified on the basis of what subfield they work in, the type of work they do, the environment they work in, and level of expertise.
- How mechanical knowledge adds to an electronics technician's expertise.
- How computer knowledge is a must for every electronics technician.
- What "soft side" knowledge an electronics technician needs.
- What knowledge and skills are required for an electronics technician to move into supervision.

The job description in Figure 3.1 was issued by a medium-size city within the Los Angeles metropolitan area. The position is for a communications technician. Look it over. Does it interest you? Is the salary appealing? Can you see yourself doing this kind of work? Is a career as an electronics technician for you?

Of course, your interest may not be in communications electronics. It may tend toward medical, consumer, automotive, or industrial electronics, to name but a few of the specialized subfields. Nonetheless, the job description in Figure 3.1 does give you some idea about the requirements and qualifications needed by an electronics technician. Note that the emphasis in the job description goes beyond the technical to such duties as writing, training, and interacting with other people. An electronics technician is not a "one-dimensional person"; he or she has many facets and must possess a variety of skills and abilities. It is a demanding, challenging, and worthwhile career, one of the few remaining that keeps you involved with things as well as symbols and allows you to feel the meaningful connection with a finished product. It could be a career for you. Hopefully, this chapter will go a long way in helping you to decide.

The chapter begins with an examination of *tech types*—that is, how technicians are classified on the basis of the field they work in, the type of work they do, where they do their work (on the bench or in the field), and their level of expertise. We see a technician at work, and we try, in a preliminary way, to see just what it is like to be a tech.

Next, we explore the *technician-plus, hard-side* factors that are important to your success as a technician. We see why you might need support knowledge in areas such as biology, automotive mechanics, or aviation technology to succeed in your chosen subspecialty. We also investigate the requirement for mechanical and computer know-how that every electronics technician must develop.

Finally, we examine the nontechnical, *technician-plus, soft-side* skills every technician should want to hone to perfection. We see why proficiency in communicating, selling, training, and supervising can enhance your value as a technician and at the same time add interest, variety, and professionalism to what you do.

Tech Types: One Kind Does Not Fit All

In 1996 (the last year for which there is data as of this writing) there were 406,000 electronics technicians in the United States. According to the U.S. Department of Labor, 141,000 worked as computer and office machine repairers, 116,000 as communications equipment mechanics, 67,000 as commercial and industrial electronic equipment repairers, 37,000 as telephone installers and repairers, and 45,000 as electronic home entertainment equipment repairers. Geographically, employment is distributed in much the same way as the population at large.

COMMUNICATIONS TECHNICIAN

Hourly Salary	$16.90; $17.80; $19.10; $20.10; $21.25
Filing Date	Last date to file applications: Feb. 5, 2001
Exam Date	Interview Date: March 1, 2001
Typical Duties	Under supervision, a communications technician installs, repairs, modifies, and maintains base station, vehicular, hand-held, and microwave radio equipment; installs and repairs specialized electronic test equipment; maintains FCC divisional records and the inventory of spare parts; prepares reports and charts; sets up temporary radio station for emergency use; and performs other tasks as assigned.
Qualifications	Requires a minimum of two years' recent experience in the installation, maintenance, repair, and troubleshooting of electronic communications equipment. Must have a General Class FCC license or current A.P.C.O. Certificate. Two-year Associate Degree in Electronics Technology from a community college or recognized technical school is extremely desirable.
Knowledge Skills Abilities	A communications technician must have a thorough knowledge of methods, procedures, and tools used in the maintenance of communications and related electrical equipment. He or she must possess the ability to read, interpret, and design sketches; maintain records and prepare reports; and train subordinates.
Vacancies	There are currently four vacancies.

Figure 3.1
Job description

As an example of what is happening, and why, in just one of those fields, let's again turn to consumer electronics. The U.S. Department of Labor projects that employment opportunities for electronics technicians in consumer electronics, including computers, is expected to increase substantially by 2005 in part because:

- The dramatic effect of switching from analog to digital technology is revolutionizing consumer electronics, creating countless new products and services that are changing the way we live and work.
- The information explosion has created a need for electronics to help us manage information and communicate more efficiently and effectively.
- Consumer electronics is a relatively stable industry, even in a recession economy; value-conscious consumers can repair the old as an alternative to buying new.
- Many consumers, tired of the high cost of movie-going and dining out, are turning their homes into entertainment centers, which can be as simple as a stereo hooked up to a TV or as elaborate as a home theater with big-screen TV, DVD player, and six speakers.
- The significant increase in home-based businesses has created a sizeable new market for computers, facsimile machines, and other home office products.
- Americans' insatiable appetite for Star-Wars™-like adult toys encourages the development of a bountiful menu of super-gadgets to feed it.

The need and demand, therefore, are clearly there. But what types of technicians will be required? What kinds are there right now? For instance, is a person a medical electronics technician, a service technician, a field tech, a supertech—or could he or she be all these? This, in turn, leads to the obvious question: How and on what basis do we classify, or categorize, technicians?

After combing the literature and questioning dozens of technicians, supervisors, and electronics instructors, I have come up with the four-part classification scheme shown in Table 3.1. It is not the only such scheme that could be developed, of course. But it does give us an outline to work with and a point of departure for further discussion. Let's examine it more closely.

There are four areas of classification. First, we can categorize electronics technicians on the basis of the subfield they work in.

You could be a **computer technician,** adjusting disc drives, aligning video monitors, installing local area and wide area networks (LANs and WANs), or maintaining servers, mainframes or super computers. Cindy Hyde is a LAN specialist for a large aerospace firm in southern California. "I'm on a LAN operation from planning to final installation and start-up. And after that I maintain them [LANs]. Even though I am not an en-

TABLE 3.1
Classifying technicians.

Industry Subfield	Type of Work	Working Environment	Level of Expertise (American Electronics Association)
Computer technician	Maintenance electronics technician	In-house bench technician	Electronics technician
Medical electronics technician	Electronics test technician	Field technician	
Broadcast technician	Installation technician		Electronics technician II
Automotive electronics technician	Electronics research technician		Electronics technician III
Communications technician	Prototype technician		Electronics technician associate (supertech)
Consumer electronics technician	Service technician		
Industrial electronics technician			
Avionics technician			

gineer, I am an A+ Certified electronics technician. I have considerable responsibility."

Or, you could be a **medical electronics technician,** repairing complex intensive-care instruments, calibrating delicate input sensors, or prototyping a new ultrasound scanner. Kathy Aims is a medical electronics technician for a medical center in Los Angeles. "I specialize in portable, liquid-flow monitors. You'd be surprised at the variety there are. It's interesting work. I like knowing I am working on equipment that impacts on patient care."

If you were a **communications technician,** you might install mobile two-way telephone equipment, set up antenna farms, check out video monitoring equipment, or test a marine radar unit. Raymond Hurd is an entry-level installation technician for Advanced Electronics in Gardena, California. "I got this job coming directly from the college intern program. I know I'm working at the bottom, for $10.50 an hour. But it's good experience. I'm involved in the installation of every type of equipment the company makes. Soon I'll be getting inside that equipment."

An **industrial electronics technician,** on the other hand, might install, adjust, and maintain a robotic cell, a programmable logic controller, an automatic guided vehicle (AGV) unit, or an entire computer integrated manufacturing (CIM) system. Ron Bagwell is an AGV specialist at the $230 million Los Angeles Times Olympic Plant. "We have 29 AGVs prowling around delivering 2000-pound rolls of paper from storage to the presses. I am in charge of keeping them moving. I handle the batteries, electrical motors, electronic sensors, microprocessor boards, and communications links—everything that makes them go is my responsibility."

In all such cases, you would be classified primarily in terms of the electronics subfield in which you work. You would be known first—but not exclusively—as a computer tech, a communications tech, an industrial tech, and so on. The electronics equipment you work on is of a specialized type, and you have specialized because you work on it.

We can also classify electronics technicians on the basis of the *type of work* they do within their subfield.

You could be a **service technician** who, for example, services, or repairs, biomedical electronics equipment. You don't design the equipment, you usually don't install it, you don't build engineering prototypes—you repair the equipment.

Or, you could be an **installation technician** (a definite growth area) within your subfield or across subfields. You might install (get up and running) the communications link to a computer-based diagnostic tool at a local hospital.

As a **maintenance electronics technician,** your main responsibility would be to keep electronics equipment operating. You wouldn't necessarily install or repair equipment, but you would provide routine maintenance, which might involve cleaning, adjusting, and replacing components, circuits, and electromechanical elements (Figure 3.2).

Or, you might be an **electronics research technician.** You would work in a laboratory right along with engineers and scientists on the development of new electronic products or services. You would build breadboard circuits, build engineering prototypes, gather sophisticated data, and even improve designs when necessary. Often your tasks would be indistinguishable from those of the engineer.

As an **electronics test technician,** your responsibilities might be to align and test assembled units as they come off a production line. At Construction Laser Systems (CLS Industries) in Hawthorne, California, test techs align laser optics, focus beams, and thoroughly test every $20,000 laser before they ship it to a customer.

Figure 3.2
A maintenance technician.
Source: Cleveland Institute of Electronics.

Finally, you could be a **prototype technician.** Your job would be to build the first working model of the engineering design. In addition to having extensive knowledge in electronics, you would be expected to be skilled in PC-board fabrication techniques, sheet-metal forming, anodizing and painting, and enclosure design.

A third level of classification we can use is your *work environment.* For the most part, you would be either an **in-house bench technician** or a **field technician.**

As the name implies, as an in-house bench technician, you work at a factory or service center. Equipment to be prototyped, aligned, tested, repaired, etc., is brought to you. Your working conditions are relatively stable, and you probably work in the same location and on the same bench every day. Every day usually means 8 hours a day, 5 days a week. You can plan a night out or a weekend ski trip with little fear that job demands will interfere with your social or family life.

A field tech, of course, goes to the equipment site, usually in a well-equipped van or, at the least, with a suitcase full of specialized tools. The equipment, or part of it, is repaired or replaced right on the spot.

As a field tech, you get out of the office a great deal. You're on the go, possibly flying to a vendor site, getting in and out of your van or automobile, meeting new people, and acting as a "white knight" to save the day by quickly getting the faulty equipment up and running.

But the field tech's job can also be quite stressful. If you are a service field tech who fixes out-of-order equipment, you are just as likely to be greeted with folded arms, tapping feet, and scowls as smiling, cheery faces, at least until you make everything right again—pronto. Also, you stay until the job is done, whether it's a late-afternoon call across the city or on a Friday out-of-town trip to Butte, Montana. Dinners out and weekend getaways are always subject to last-minute cancellation.

Finally, you can be classified as an electronics technician on the basis of *level of expertise,* as determined by various accrediting agencies, such as the American Electronics Association.

You might be an **electronics technician I,** a person who performs a variety of routine tasks to assist engineers with electronic systems. Requirements are the ability to use electronic theory and basic machine tools, plus the ability to work from schematics and rough sketches in completing assignments.

Or, eventually, you could become an **electronics technician associate (supertech).** You would perform a wide variety of complex technical tasks to assist design engineers with developmental operations. Requirements are a minimum of 6 to 8 years related experience with a 2-year technical degree.

Of course, as we have suggested all along, your occupation as an electronics technician could cross all four technician categories. As illustrated in Table 3.1 with a shaded line, you may well be a *medical electronics technician, servicing* biomedical equipment in the *field* at an *electronics technician I* level of expertise.

José Hernadez is a good example of what we mean. Twenty-nine years old, he is a graduate of Truckee Meadows Community College in Reno, Nevada. He started working as an electronics level I technician at Taski Medical Instruments, Inc. right after completing an associate of science degree. He has been with them ever since.

In a typical day, José is "on call," out in the field, to various local hospitals as a service technician for Taski. When he arrives at a hospital and surveys a faulty piece of Taski medical equipment, his first task is to determine how long it will take to get the equipment fully functional again. Down time is critical and must be kept to a minimum. If the problem is more than mechanical or beyond the incorrectly hooked up, out-of-adjustment level, chances are it's a component failure on a PC board. José must quickly determine which board is at fault and then decide either to replace the entire board or to troubleshoot and repair down to the component level, right there in the hospital. In nine out of ten cases he will elect to replace the entire board at once. Remember, the goal is to get the equipment up and running as fast as possible. The faulty board is then returned to the Taski service center in Houston, Texas, where an in-house bench tech troubleshoots it at the component level and produces a rebuilt, fully tested board, ready to be installed again in Taski equipment.

Thus, José can be considered an entry-level biomedical service technician working in the field. Since he already has his A.S. degree, he is definitely on his way to becoming a supertech.

Technician Plus: More Than Electronics Know-How Is Needed

Experienced electronics technicians, whether employed in small, medium, or large companies, whether working as biomedical or aviation technicians, as service or maintenance techs, on the bench or in the field, or at various levels of expertise, understand that knowledge of electronics and skill in applying it are the cornerstones of what they do—and what they get paid for. But they are also aware that "doing electronics," being handy with a soldering iron, and being an expert troubleshooter, are generally not enough. Their job often requires that they have knowledge, training, aptitudes, and skills that go beyond electronics know-how.

These *technician-plus factors,* as we might call them, may be broadly grouped as *hard side* and *soft side.* **Hard-side** factors refer to additional technical expertise that may be needed, such as a knowledge of biology for a medical electronics technician or the computer skills required by all technicians. **Soft-side** attributes are in support areas, such as communications, selling, training, and supervisory skills. These technician-plus factors are essential job prerequisites in many cases. As such, they warrant further discussion.

Let's examine the hard-side factors first. At one level, the need is obvious. Take the biomedical technician, for instance. He or she must have an understanding of biology in addition to a knowledge of electronics. The curriculum of a biomedical electronics technician contains required courses in biology, anatomy, and physiology. As Frank R. Painter of NovaMed, a medical equipment service company, says of his technicians: "Our people understand anatomy and physiology as well as electronics. They speak the same language as doctors and nurses." Thus a biomedical electronics technician has a core hard-side background in a subject area other than electronics.

Consider an automotive electronics technician. Obviously such a person must know automobiles in addition to electronics. Indeed, most such techs were auto mechanics first and moved on to specialize in automotive electronics. But given the increasing complexity of the latter, it is likely that more and more automotive electronics technicians will be electronics technicians first and auto mechanics second. The day is not far off when every automobile dealership will require a full-time electronics technician. He or she will have to know how to use a monkey wrench as well as a multimeter. If your main interest is electronics but you like cars, too, you're going to have to think like a mechanic as well as an electronics technician.

Then there is the field of avionics. The word alone tells you that people entering this subspecialty with the intent of doing electronics had better have aviation on their minds as well. A thorough knowledge of airframe mechanics is just as important as a knowledge of electronics. As Jack Gear, of Aerospace Corporation, Los Angeles, California, says: "While you are expected to know an amplifier circuit from a power supply, you'll also need to distinguish an aileron from a pylon."

The foregoing examples in medical and automotive electronics as well as avionics illustrate the subdisciplines that are often defined in an electronics subfield. In each case, the subdiscipline is usually common to a particular branch of electronics. In other words, a knowledge of biology is required for medical electronics but not for automotive electronics. However, there are hard-side nonelectronics skills of a more general nature that every electronics technician would be advised to acquire, regardless of his or her specialty. Two come quickly to mind: *mechanical know-how* and *computer know-how*. Let's take each in turn.

Electronics technicians need to have some facility with mechanical things. They must know the difference between a flat-head and a Phillips screwdriver, and they need to feel comfortable working with gears, pulleys, and a host of electromechanical components. During a day's work as a tech, they will be called upon to use such mechanical know-how in a variety of ways.

First, just to remove, disassemble, and put back together much of today's electronic equipment demands mechanical aptitude. An electronics technician can easily spend as much time removing a panel-mounted, octopus-cabled, high-frequency oscillator, painstakingly opening the unit, and eventually returning it to its rightful place as he or she will in actually working on the electronics inside.

Second, much of what needs to be repaired by electronics technicians is of a mechanical or electromechanical nature. Videocassette recorders (VCRs) are notorious in this regard. It is estimated that VCR repair is 90 percent mechanical (belts, pulleys, gears, cleaning, and so forth) and only 10 percent electronics. The same is true for photocopier equipment. James Hunt, a technician at Xerox Corp., said it best: "If it's an electronics problem, I don't even mess with it here [on site]. I take the circuit board back to the service center. I was hired as an electronics technician, but in the field I am really an office-machine mechanic."

A third reason for boning up on your mechanical skills is a bit more self-serving. Many companies give prospective job-seekers an entrance exam. Such exams are often filled with questions on levers, gear ratios, and various mechanical procedures. You'll want to pass those exams. Knowing a fulcrum from a spindle and a piston from a valve will help.

In sum, when it comes to mechanical know-how, the bottom line is clear—as an electronics technician, you can never be too knowledgeable or experienced in things mechanical.

We can say the same thing about computer know-how—both hardware and software. We refer here not only to a knowledge of data-processing computer systems, such as PCs, but to process-control computers, or microcontrollers, embedded in the electronic devices they control. With reference to the latter, as an electronics technician you will encounter an ever-increasing array of electronics products that are microprocessor-controlled.

Some will experience hardware problems, but for many more the failure will be in the software. You'll need to know about that software, how it is structured, and how programs are written in assembly language. Furthermore, in a number of cases, what distinguishes one electronics control device from another is not hardware at all. In that respect, the devices are very much alike; what is distinctive is the unique program each one is running. Your repair work will involve manipulating the program, or software.

Then, of course, there are the personal computers, workstations, and servers you will confront in every business environment you enter. Even if you're not a computer tech, as an electronics technician, you will undoubtedly be seen as the "expert on the scene" when trouble develops. Often you will be able to avoid involvement by claiming ignorance of the specialized equipment or by pointing out that such equipment is under warranty or a service contract. Many times, however, you will not be able to duck so easily. You'll have to dig in and get the machine operating again. As Fred Blechman, author of *Simple Low-Cost Electronics Projects,* told me: "The public thinks if you know anything about electronics you know everything about electronics."

With regard to computer know-how, remember this: Almost everything you touch in electronics is computer-based. As an electronics technician, no matter what type, learn all you can, whenever and wherever you can, about computers—all aspects, hardware and software.

Soft-side, in contrast to hard-side, skills are more subtle and diffused. The abilities to *communicate* ideas and information, help *sell* products and services, *train* fellow technicians, or *supervise* employees fall under a broad heading known as *people skills.* If you are a technician with ambition and hope for advancement or, at the least, with a wish to make your workday more interesting, varied, and purposeful, you will want to expand your expertise in soft-side skills. Let's examine more closely what this means.

In any technical field, and most certainly electronics, **communications skills** are essential. Through speaking, listening, writing, and reading, important information and ideas are exchanged.

In explaining to a customer your company's repair policy, in dividing up the work schedule with fellow employees, or in presenting a new production plan to your supervisor, you'll want to speak as clearly, precisely, and persuasively as possible. What you say as well as how and when you say it will impact strongly on your success as an electronics technician.

Listening skills, though often given less attention, are just as important as—or maybe more important than—speaking skills. As a technician, you must learn to listen with real interest and empathy. Through questioning, paraphrasing what a speaker has said, and providing nonverbal feedback—nodding or shaking your head, for example—you show fellow workers, customers, or supervisors that you honestly value their comments.

Speaking and listening occur in real time, of course. To go beyond the immediate, in time and place, we must write and read. In documenting what you have been doing, be it in a laboratory or at a customer worksite, clear and concise written technical expression is a must. In some cases what you have written becomes a legal record, to be used in adjudicating disputes as simple as deciding what equipment is to be repaired next or as vital as determining who gets credit for a major new invention.

Above all, as an electronics technician you must read—quickly, critically, and with comprehension. The solution to many, if not most, of the technical problems you encounter will be found in the written word—in data manuals, operation manuals, repair manuals, memos, articles, books, and websites. The willingness to read these documents and the ability to read them effectively will affect your success as an electronics technician just as much as your skills with a soldering iron or an oscilloscope.

Communications skills are not something you are born with; they can be acquired like any other skill, such as tracking down a short circuit or debugging a computer program, with practice and commitment. Through interaction with fellow students, employees, and friends and with the work you do in the classroom or laboratory, you can begin to develop and enhance your speaking, listening, writing, and reading skills. And when you do, be particularly aware of cross-cultural cues. As Dr. Joseph Tajnai, a Productivity Manager for Hewlett-Packard, explains: "With the global economy as it is, our techs communicate with counterparts and supervisors all over the world, from Malaysia to Italy. They need to be aware of national cultural characteristics."

With strong communications skills, especially in speaking and listening, you're ready to lend a hand in **selling** the electronic products and services your company provides. "Selling? If I wanted to be a salesperson, I would have majored in business or skipped college altogether. I thought being an electronics technician would place me as far away from sales as I could get." If that's you talking or thinking, then think again. As an electronics technician, if you are in contact with customers, whether by installing, maintaining, or servicing their equipment, what you say and do can have a significant impact on company sales and, ultimately, on the bottom line.

Customers usually want to know what new products and services are available, and as the technician on the scene, you are the one to whom they will frequently turn for information and advice. Furthermore, you are often in the best position to suggest when the current equipment is beyond repair or has simply outlived its usefulness. A few thoughtful suggestions from you, a technical person who is seen as relatively objective, can go a long way toward helping your company make a sale.

Actually, some companies expect their electronics technicians to help sell products as a part of their job. Larry Ornstein, of Marine Electronics in Hermosa Beach, California, is really a part-time salesperson while being a full-time technician: "When I go out on a boat to install or repair our equipment, I am expected to keep the sale of new products firmly in mind. It isn't that I go out there as a pushy, high-pressure salesperson; I'm not that sort of guy anyway. My main job is technical, installing and repairing. But if I think the customer [boat owner] would be interested in an update or something new to expand his [or her] marine electronics inventory, I wouldn't hesitate to discuss the matter with him [or her]. My boss expects me to, and I am actually happy to do it. It makes my job more exciting. Besides, if what I suggest results in a sale, I get a small commission."

A firm foundation in communication skills can also lead to facility in **training** or instructing. Throughout your workday, you will often be called upon to use such skills to train or instruct fellow employees or customers.

In the job description at the beginning of this chapter, note that an applicant must possess the "ability to train subordinates." Much of what you learn as a new technician will come from the more experienced techs working around you. As in a buddy system, you'll be placed in a one-on-one situation to be tutored, instructed, and advised. Sooner than you may think, you will be asked to do the same thing for someone else. You will be the teacher as well as the student.

The same sort of thing is true with regard to customers. Someone has to show them how to operate and maintain new equipment. For example, a high percentage of the troubles reported in electronics equipment are due to the lack of training on the part of operators. The technician will have to show the operator the problem without hurting his or her feelings. The technician will have to be sensitive to the customer's needs. In short, the technician will have to have empathy. Yes, in some instances vendor training is available. But this isn't always the case. Much of the short-term, immediate instructing is left to the technician.

How do you develop such training, instructing, or teaching skills? As mentioned earlier, communications skills, especially speaking and listening, are fundamental. Beyond that, there is much you can do right now, while in school, to practice and sharpen such skills. For example, in a laboratory setting, you can learn how to help fellow students. You must be careful not to give too much information or do the work for your colleague. But by listening to find out the exact problem and then formulating and articulating a solution, you will develop your instructing and training abilities.

Also, you might volunteer as a tutor in your department or school. You don't have to be the best there is at your subject to make a good tutor. All that's needed is patience, empathy, and a willingness and ability to explain material in straightforward terms.

There is a bonus to all this instructing, training, and tutoring, by the way. As any teacher knows, you learn a great deal about a subject while teaching it. This is true because teaching forces you to analyze the material, synthesize it, eliminate the superfluous, and focus on the main points in a logical and coherent manner. The bottom line is that by instructing others, you'll gain just as much as they do—maybe even more.

Finally, as your technical proficiency matures and your communications, selling, and training skills improve, you may be asked to take on supervisory duties. A **supervisor,** as distinct from a manager, is expected to have strong technical know-how and be an example of the best in the field. A supervisor often continues to perform technical duties while supervising and providing guidance to others doing the same sort of work. A supervisor, more than anything else, must have the respect of subordinates. He or she must present an example technicians wish to emulate. While remaining technologically strong, the supervisor must maintain keen people skills, for it is often these skills that have been responsible for the advance to a major new responsibility. The supervisor is the best example of what is meant by technician plus—a technically competent individual with hard-side and soft-side areas of expertise.

All this may seem a bit overwhelming. To begin with, an electronics technician is expected to know electronics—that is, to be able to install, maintain, test, or service complex electronic equipment. He or she must understand how to operate sophisticated stand-alone and computer-based test instruments and, in some cases, build prototype products from scratch.

In addition, there are the hard-side issues. Specific training in biology, automotive mechanics, aviation technology, and so forth may be required for certain subfields within electronics. And, of course, there is the question of mechanical and computer know-how—more technical expertise to gain and maintain.

But that's not all. There are the soft-side skills in communications, selling, training, and perhaps supervising that must be considered and incorporated as needed. We weren't kidding when we said earlier that an electronics technician is not a one-dimensional person.

Yet, if it were easy to succeed as an electronics technician, everyone would be doing it. Everyone is not. As a result, the demand for qualified technicians is strong. If what has been discussed in this chapter is seen as a challenge rather than an insurmountable hurdle, the rewards can be considerable, not only financially, but in terms of self-esteem and respect of colleagues in the work force.

Summary

In Chapter 3, we learned what the demand will be for electronics technicians in the next few years. We saw how electronics technicians are classified in terms of the subfield they work in, the type of work they do, their work environment, and their level of expertise. We examined hard-side skills, such as mechanical and computer know-how. And we did the same for soft-side skills, those involving communications, selling, training, and supervision. We concluded that when it comes to an electronics technician in the new millennium, one size does not fit all.

Review Questions

1. The U.S Department of Labor expects the job opportunities for electronics technicians to increase _____ by the year 2005.

2. If you are identified as a computer technician, a medical electronics technician, or a communications technician, you are classified primarily according to the _____ you work in.

3. If you are identified as a service technician, installation technician, or maintenance electronics technician, you are classified as to the _____ of work you do.

4. An in-house, bench technician or a field technician would be classified according to his or her work_____.

5. An electronics technician I, II, etc. is classified according to level of _____.

6. Mechanical know-how and computer know-how are _____ -side factors in referring to additional technical expertise.

7. The abilities to communicate ideas and information, help sell products and services, train fellow technicians, or supervise employees fall under a broad heading known as _____ skills.

8. With strong communications skills, especially in speaking and listening, you might be ready to _____ electronic products and services.

9. A firm foundation in communications skills can lead to facility in _____ or instructing.

10. A _____, as distinct from a manager, is expected to have strong technical know-how and be an example of the best in the field.

Individual or Group Activities

1. Select a piece of electronic equipment you own (stereo, electronic game, calculator, CD player, VCR, etc.). Write a set of operating instructions for the basic features of the device. Assume that the reader has no technical background. Further assume that the instructions will be used by someone with the equipment present.

2. Give a 3- to 5-minute oral presentation on how to operate a piece of electronic equipment such as a stereo, CD player, or VCR. Be sure your talk includes a discussion of what the equipment does, any safety precautions to observe, and what can happen if instructions are not accurately followed.

3. Obtain a half-dozen job descriptions for the occupation of electronics technician. To do so, contact the personnel department of various public agencies and private companies. Compare the job descriptions in regard to salary, qualifications, and duties required. Write a 1- or 2-page summary comparing the data you have acquired.

4. Interview an electronics technician who has gone into technical sales. How important was a technical background in the individual's success? Why did he or she choose to go into technical sales? What people skills contribute to the salesperson's success? Present your results to the class in an oral report.

5. Interview an electronics technician who has gone into technical writing. Why did the technician make the transition? What are the pros and cons of being a technical writer? Does the technical writer spend most of the time writing or simply handling documents? Present your results to the class in an oral report.

6. Interview an electronics technician who has gone into corporate training. What attributes have contributed to his or her success? How does corporate training differ from high-school or college teaching? Is a bachelor's degree required? Present your results to the class in an oral report.

Issues for Class Discussion

1. Discuss the role of the supervisor. While he or she must have the respect of his or her subordinates, the supervisor must remember that he or she is no longer one of the workers. What does this mean? Is a supervisor part of management? How is his or her role viewed by management?

2. Discuss at least a half-dozen ways that one can informally (not through class work) increase mechanical ability or know-how. Are men more likely to get such informal exposure than women? If they are, what can be done about decreasing this disparity?

3. Discuss ways one can improve or expand computer know-how. Should one take courses, or should one just "dig in" and learn by doing?

4. Compare and contrast the role of the in-house bench technician with that of the field technician. Are there different personality traits required for each? Do more experienced techs tend to gravitate to one versus the other? Is there a salary difference? Is there a status issue?

Electronics Technician Profiles:
Meet Ron, David, Kay, Gideon, and Eric

Objectives

In this chapter you will learn:

- What it's like to be an industrial electronics technician.
- How important mechanical ability is to your success as an electronics technician.
- How writing skills impact your job as an electronics technician.
- What it's like for a communications technician in a large transit authority.
- What it's like to be an A+ certified computer service technician.
- What it takes to make the transition from work as a technician to a job as manager.
- What "service" in service technician is all about.
- The ups and downs of running your own electronics business.

In Chapter 1, "The Electronics Industry: In the Air and Everywhere," we saw how large, dynamic, and diversified the world-wide electronics industry is. We examined a bit of its history, and we looked ahead to the next few years.

In Chapter 2, "Electronics: Field of Dreams," we examined the variety of technical careers open to those seeking challenge, growth, and good income in their chosen occupation. We also explored the so-called workplace revolution and saw how changes in the working environment and work force will affect employment prospects.

And in Chapter 3, "The Electronics Technician: Bringing Electronics to Life," we zeroed in on the career of the electronics technician. We surveyed the various occupational subcategories and we looked at how more than electronics know-how is needed to succeed in this challenging field.

Finally, in this concluding chapter of Part I, "Being an Electronics Technician," it is appropriate to talk with a few electronics technicians and to hear in their own words what it's really like in the real world. In doing so, we have selected five representative electronics technicians to profile, or interview.

Ron Bagwell, the first technician you'll meet, is an in-house industrial electronics technician working in the exciting area of automated control. In listening to him talk, you'll get a better idea of what it's like to work as a technician in a modern industrial plant.

David Wells is an A+ certified computer service technician. You'll discover how he has combined technical know-how with a management background to succeed in the competitive computer equipment industry.

Kay Koopman, as you will see, is a communications technician working for a large public employer, the Southern California Rapid Transit District. As you get to know her, you'll discover what a typical day in the life of a field tech is all about.

Gideon Green, a field service representative, not only fixes high-end fax machines, he has made the transition to management. We'll see what he has to say about what it takes to work in both worlds.

Finally, Eric Chavez is a technician who has gone into business for himself. After digesting his comments, you'll understand better the rewards and pitfalls awaiting anyone willing to undertake this, the most momentous of challenges.

Ron Bagwell: AGV Electronics Technician

Ron Bagwell is a thoughtful, 32-year-old, level 2 industrial electronics technician responsible for the maintenance and servicing of 37 automatic guided vehicles (AGVs) at the $230 million Los Angeles Times Olympic printing plant. The factory can put out 50,000 newspapers an hour on each of its six gigantic four-story color presses. The 684,000-square-foot plant is a state-of-the-art enterprise, the hub of which more closely resembles the flight deck of a spaceship than the clanking, locomotive-like, cast-iron behemoth your mind might now be conjuring up (Figure 4.1). Every aspect of newsprint production, from the off-loading of 1-ton rolls of paper to the final sorting and bundling of printed and stuffed newspapers, is controlled electronically. The plant is arguably the most highly automated printing facility in the world today—thanks entirely to modern electronics.

The following interview between Ron (Figure 4.2) and the author took place at the Los Angeles Times Olympic plant.

Ron, what is your job classification and what are your duties?

I am an electronics technician. I, along with my fellow technician, Terry, am responsible for the maintenance and repair of automatic guided vehicles, or AGVs, as they are called.

What's an AGV?

These are battery-powered vehicles that travel at approximately 180 feet per minute along a guidepath, following an imbedded wire network which emits five different frequencies. Transponder cards located at decision points

Figure 4.1
Los Angeles Times Olympic printing plant.
Source: Ronald Reis.

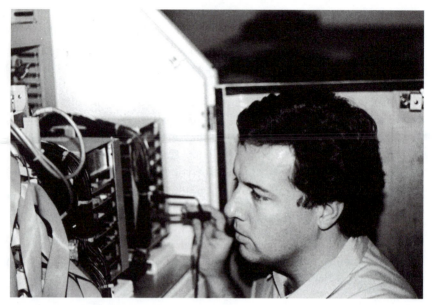

Figure 4.2
Ron Bagwell, industrial electronics technician.

along the guidepath contain function codes. As an AGV passes over a transponder card, it performs functions based on what it reads from the transponder card and what the computer (system controller) requests it to perform. Twenty-nine HV-3 AGVs are used to transport rolls of newsprint to the presses. Eight tugger AGVs are used to transport newsprint throughout the warehouse for storage.

They sound quite complicated.

They are. When you consider all the radio-controlled circuitry, the optics, the guidance and safety circuitry, plus the mechanics—they are a moving, lifting vehicle, remember—yeah, they are fairly complex.

And you and Terry are responsible for all 37?

Yes. When I arrive in the morning, the first thing I do is check out the disabled vehicles, all of which have been parked in our repair and maintenance area.

What's usually wrong with them?

It varies. In many cases the same problems appear over and over again: it's [the vehicle is] losing guidance, communication problems, vehicle not reading transponder cards, optical problems, and so forth. On the other hand, some have intermittent electrical problems, which are always hard to fix.

What about mechanical problems?

Sure, that's part of it. Due to an engineering flaw we were having problems with the drive motor. Eventually we had to replace the drive units on all the vehicles.

So, having mechanical ability is a plus?

It's more than a plus, it's a necessity. I don't think I would have been hired for this job without being able to demonstrate mechanical aptitude.

Speaking of getting the job, let's talk about your background for a moment. How long have you been here at the Times?

I started here when I was 18, right after graduating from high school.

As an electronics technician?

Oh no! I started as a pusher and shover. I pushed rolls of paper around on a dolly to the various presses. Then I shoved them onto a pickup spindle.

Like what the AGVs do today?

Exactly, I now maintain and service the very machines that replaced me.

That's fascinating. You've turned the automation dilemma on its head, so to speak?

Yeah, I am the master in the end.

When did you get interested in electronics?

A few years ago, I decided I needed more training if I was going to get ahead. I had a friend who was going to an ITT Technical Institute. He was excited about it, that's all he could talk about. I decided to check out the various schools, both community colleges and private schools. Finally, I settled on the DeVry Institute of Technology in the City of Industry. I went there for 2 years in the evenings.

Why did you choose DeVry?

I liked their night program of 3 nights a week. DeVry was very accommodating in terms of schedule.

Were you under a grant, loan, or scholarship?

No. I paid for it all myself.

Did you graduate from DeVry?

Yes, I have a digital electronics technician certificate.

So, when did you start here as an electronics technician?

When I finished DeVry, I started looking at various companies for a job. But I was already making good money at the LA Times compared to what I would have had to start out with as a new tech. So I decided to stay here, doing the same unskilled job, but I let everyone know I wanted to get into electronics.

You were looking to make a lateral move?

Yes. And when this new plant opened up, I came here as an electronics technician.

But you didn't know anything about AGVs, did you?

No, not at all. But Terry, the guy I work with, did. He was hired away from the company that sold us the AGVs. He's a great teacher—mentor, actually. He has taught me a great deal.

What do you like best about your job?

Being involved in the latest technology, like robotics.

Robotics?

Yes. I consider AGVs to be robots. And now that they are out of warranty, we have to repair them down to the component level. We used to send the bad boards back to the company. Now we are responsible. I like that because I am learning a lot more. It's challenging.

What do you dislike about your job?

I really have no complaints. I like working with something tangible. I like seeing results. When an AGV comes in, it's broken. When it leaves here it is fixed. Terry and I make that happen.

Do you and Terry get along?

Sure! We work well together. I get along with everyone here.

Is that important? Getting along with co-workers?

Of course. We are all part of the same team.

What about personal communication skills? Do you have to do any writing?

A little. When we open or close a job, they want us to write a few sentences. You have to explain what you did to resolve the problem.

What, if any, computer skills do you have to have in your job?

All the AGVs are controlled by a central microcomputer. I have to know how to operate that computer, you know, boot it up, DOS keyboard commands, and so forth. I definitely have to be computer literate.

Would you recommend this career to a friend?

Sure. I get a lot of satisfaction from my work. My friends and family are proud of what I do.

What do you see for yourself 5 years down the road?

I don't know for sure. I have begun to fix my friends' VCRs, TVs, and so on. I haven't charged them, but I might begin to. That way I could earn extra money and gain more experience in electronics.

Have you ever thought of supervision or management?

That does interest me. I am going back to DeVry now to finish my A.S. degree. After that I might go to a 4-year college and get my B.S. in industrial technology. I haven't decided on the latter though.

Well, good luck to you in whatever you choose to do. And thanks for taking the time to talk with me.

Thank you for asking me.

Ron, as you see, is enjoying a challenging career as an electronics technician. At least three points are worth noting with regard to his interview.

First, did you notice Ron's comments about mechanical ability? He and Terry are responsible for the service and maintenance of the entire vehicle, not just its electronics. That requires knowledge, familiarity, and experience with things mechanical.

Second, Ron's comments about automation and how he now works on the very vehicles that replaced him in his earlier job were interesting. He didn't let automation eliminate his job—he moved into a new job where he controls the automated vehicles.

Third, note that Ron made a lateral employment move—that is, from one job category to another within the same company. You may find yourself in a position to do the same thing. If you are already working part- or full-time, doing routine, unskilled or semiskilled work in a nonelectronics capacity, it may be to your advantage to look around your company and see what is available in electronics. Many companies would prefer to move you into such a position than hire someone from the outside. And, of course, the advantages to you can be considerable: familiarity with company policy and personnel, continued benefits, most likely a favorable salary position, seniority, and so forth. Even if you are working for Federal Express as a shipper, Hughes Aircraft as a stockperson, or a rapid-transit district as a dispatcher, look around. All these companies need electronics technicians. Let everyone know of your interest in moving into electronics. Get the word out; that's the point.

David Wells: A+ Certified

When Electrorent, an 800-person corporation that rents and leases computer equipment, from laptops to Sun servers, went looking for someone to manage their equipment services group in Van Nuys, California, they found the ideal candidate in David Wells (Figure 4.3). With 25 years experience in aerospace, most of it in management, and a recently acquired A+ Certification in hand, David, at 54, possessed the supervisory background, technical expertise, and personal maturity the position would require. Six months on the job (as of this writing), David manages a group of technicians (80 percent of whom are also A+ certified). They are responsible for the testing, repairing, and configuring of all data processing equipment that leaves the warehouse.

I interviewed David at Los Angeles Valley College, the community college where, under a JTPA (Job Training Partnership Act) program he had, a year earlier, completed a 600-hour course to prepare him for the A+ certification examination.

Figure 4.3
David Wells, A+ certified technician

You've been around awhile, David, what is your background?

After graduating from high school, I bummed around a bit before deciding to join the navy. That was back in 1962. It was great. I traveled the world and even got to see an A-bomb blast on Christmas Island in the Pacific. Since I'm still here, I guess the radiation damage was minimal.

Were you doing electronics in the navy?

No, I was a machinist mate for 2 years. It would be some time before I got interested in electronics.

What did you do after being discharged?

I found a job at Rocketdyne, an aerospace company that built the Saturn Rocket engines. Then, it was on to Lockheed, where I found a niche in cabin interiors. I worked my way up to division manager in charge of interiors for the L-1011.

You reached that level of management without a college degree?

Yes, in those days having a degree in business wasn't critical.

So, when did you get interested in electronics?

In the mid-80s I worked on the B2 stealth bomber program at Nothrup. Much of what I did is classified, so I can't talk about it. But I will say this, it was electronics in nature. But in 1996, the program was canceled. I was 50 years-old and out of a job.

What did you do then?

Even though I had a background in mainframe computing, I knew in today's world I needed to bring myself up to speed on PCs and Microsoft products. So, I found the JTPA program and went back to school at 51. Yes, there is life after 50.

Were you studying to pass the A+ Exam in the JTPA program?

Yes. It was a 600-hour course. There were 14 of us in the class. To my knowledge, all wound up A+ certified.

Was the course tough?

It was intense. We began with hardware, you know, tearing down PCs, putting them back, reconfiguring them. We also messed with printers and other peripheral devices. Then it was software, Wiondows 95, DOS. We even did a little MCP as part of the program.

What's MCP?

It stands for Microsoft Certified Professional. It is another certification worth having.

Tell me about the A+ Examination?

There is the core test and the advanced test. I took the core test first. It lasted 90 minutes with 90 questions. It is all computer-based testing. You know right away if you passed. I did. Then, a few weeks later I took the software portion. That was 90 questions, too. I passed that, also.

What's the test environment like?

You are basically partitioned into a room by yourself. There is a camera in the room, keeping an eye on you. You will be monitored as to what materials you can bring in. No calculators are allowed, but you won't need one anyway.

What suggestions do you have for those wanting to take the test?

There are lots of test preparation sites on the Internet to go to. They will provide you with tons of questions. There are also chat room-type sites where those who took the test will put in their comments. I recommend cruising the Internet for such information.

Does the test change over time?

You bet! Now they have what is known as "active testing." If you answser the first 20 questions right, that's it, you are done, you don't have to go further. If not, then the software analyzes how you have been answering questions so far and gears the remaining questions accordingly. If you see the 21st question, you know you didn't get the first 20 right.

How did you get your first job after passing the examination?

I found Electrorent on the Internet and sent them my resume. They were looking for someone who had management experience plus a technical background. Since I did not have a degree, my A+ certification looked pretty good to them.

What's your work day like?

When equipment comes back into my area, it is first tested to see if everything is there and in working order. Some of the 4000 pieces of equipment we loaned out during the Democratic Convention came back with problems, including outright theft of hard drives, memory, and even CPUs.

And, of course, when equipment is returned from a defense contractor, the hard drive is missing. They keep it for security reasons. Obviously we must replace the drive.

If something is wrong, and it usually is 20 percent of the time, the equipment goes to repair. They do whatever it takes to get it operational. Then they load new software.

What then?

Then it is on to the configuration group. We can put together a system to meet anyone's needs. For example, a company will have a server go down.

We can ship a new one, completely configured, in 24 hours, sometimes even the same day.

Do you have techs in all three areas?

Yes. We start them out in testing. Then it is on to repair. Finally, configuration. The money follows, too. The lowest pay is in testing, the highest in configuration. Actually, though, I'm trying to cross-train my techs. That way, if on a given day I need more people in repair, I can pull them from configuration or testing.

What is the pay like?

It is $12 to $15 an hour to start. It can go as high as $25 to $26 an hour in configuration. Of course, if you move into supervision and management, it is even better.

How important is it to be A+ certified where you work?

I can't see technicians getting a job in this field without it. Even though half of our techs have an associate degree or higher, it is their A+ certification that did it for them, got them in the door.

Dave proves the cliche, "It's never too late to go back to school." Yet, notice, he didn't start over from ground zero. That is, he built on his background in aerospace and as a manager. David got the job at Electrorent because he had both technical skills and management expertise. Look to your own background. How can you incorporate your own varied experiences into the technical world of electronics? You may be surprised at how related expertise, from art to zoology, can add to your resume and make you more employable as a technician.

Kay Koopman: Electronics Communications Field Technician

Kay Koopman is a bright, articulate, 32-year-old electronics communications technician working for the Southern California Rapid Transit District (SCRTD). The SCRTD is an 8800-person, $600 million operation encompassing over 1400 square miles in the greater Los Angeles area. The district maintains a huge fleet of more than 2400 buses, or coaches, that ply the surface streets and freeways of LA on a 24-hour, 7-day-a-week basis. Each coach contains a 20-pound two-way radio, an intercom, an electronic head sign, and an electronic fare box. It is Kay's job, along with 50 other electronics technicians, to install, maintain, and service the electronic devices that fill these coaches.

Figure 4.4
Kay Koopman, communications technician.

At the time of our interview, Kay (Figure 4.4) was 6 months pregnant and on "light-duty" disability. The interview took place in an office at SCRTD headquarters.

Kay, what do you do here?

I am an electronics communications field technician. I work on the two-way radios, electronic head signs, public-address systems, and electronic fare boxes in the coaches. The new fare boxes require a lot of upkeep.

Why's that?

They're new, only 3 or 4 years old. They are all electronic, as opposed to the totally mechanical ones that went before them. They're reliable but still need a fair amount of work.

What's the advantage of using electronic fare boxes over the mechanical ones?

There are a lot of advantages. They have electronic locks, so they are more theft-proof. The money is counted electronically as it goes in. We know what tokens are being used and where. The driver has practically nothing to do with it. And at the end of the day, we can plug the unit into a computer and download all its data.

How long have you been at the SCRTD?

Just 2 years.

What did you do when you first arrived?

They had me working on the bench for several weeks. Even now, when we have time, field techs spend some time on the bench. When we do, we troubleshoot down to the board level, rarely the component level.

What then?

I moved into installation of two-way radios in the coaches. After that, it was into field service, which is what I do now. Mainly I look for problems.

What's a typical day like for you?

Ok. I show up at the shop at 7:30 A.M. The district is very concerned about punctuality. Everyone punches in beforehand. We then have a short meeting with our supervisor. Then our lead person usually assigns our duties. I'm told what division to go to that day. The district has 12 operating divisions.

Do you drive there, to the divisions?

Yes, I drive to one of the seven divisions that my group services. I have my van filled with tools and spare parts. When I get to the yard, I pick up the "B. O. [bad operation] cards." The explanations as to what is wrong are usually not very specific. You know, "bad radio," "bad fare box." Remember, it is usually the bus driver filling out the form. I then find the first coach to work on and get right to it.

What is a typical problem that you encounter?

Bad connections. There is a lot of cabling. Most of the time it's a bad connection.

What percent of calls result in your having to bring back a unit for work on the bench?

About 10 percent, that's all.

How many coaches will you work on in a typical day?

It averages 10 or 11.

That many?

Yes, maybe even more on a bad day.

Are the more skilled and experienced technicians in the field or on the bench?

People who have worked here a long time generally try to get out of the field and onto the bench because it does get a bit monotonous in the field.

Is there a night shift?

Oh sure! We work days, nights, and weekends. I am not exactly thrilled about the hours. But we have got to keep those buses rolling.

Let's shift gears a moment. Tell me about your training prior to coming to the SCRTD, if you will.

All my training in electronics took place in school, college. Before coming here I had worked in the building trades, carpentry, some electrical. But all my electronics I learned at Los Angeles Trade Technical Community College in their 2-year program, specializing in radio communications.

Do you have a degree?

Yes, I have an A.S. degree.

When you first came here, directly out of college, were you a bit apprehensive? I mean, upon seeing all that goes on around here technically, weren't you a bit intimidated?

Actually, no. They were very supportive, the technicians and the supervisors.

Does the SCRTD offer a formal training program for new technicians?

They do now. It's quite extensive. But once you're out in the field, you are kind of on your own.

What about your future prospects, promotions, etc.?

I have just received a promotion to rail electronic communications inspector, which will become effective as soon as I return from my maternity leave. The job will entail working on the new Metrorail underground. I'll be inspecting all the new electronic equipment that subcontractors are installing. Later, as the contractors pull out, I will be servicing that equipment.

What about management as a career possibility?

Yes, that's an eventual possibility.

But you like working as a field tech right now, don't you?

Sure, but it does have its disadvantages. It's hot, grimy, and dirty at times. I don't mind though.

Do you have to be a mechanic too?

Not in the sense that I work on the coach engines. But you do have to have mechanical aptitude. Some techs come in with it, others have to pick it up on the job.

Has the SCRTD been accommodating with regard to your pregnancy?

Yes. My job is guaranteed for up to a year with no pay.

How is it being a woman in an occupation that is traditionally dominated by males?

It's actually been very good, especially compared to the building trades.

What percentage of the techs are females, would you say?

I'd say 15 to 20 percent. Not too bad.

What about salary?

Salaries here are very good. That's why people stay.

How did you first get interested in electronics?

Actually, my back was up against the wall. We moved to LA for reasons beyond my control. I left all my support systems behind. I was an artist at the time. I actually did electronic art. I discovered, though, that I couldn't make a living at it. I cleaned houses for several years. Electronics seemed interesting, both the hands-on part and intellectually. So that's what I got into.

A few final points, if I may. What about the need for communications skills? Could you comment on that?

It's definitely a plus. There is a noticeable difference between the techs that have good communications skills and those that do not.

What about computer literacy?

Definitely. When I am relief lead I enter all jobs onto a computer.

Do you have any final thoughts you would like to share?

Tell your students that above all they must enjoy, even love, electronics. If they do that, then they will succeed.

Well thank you very much, Kay, for your time. Best of luck to you—and to the new baby.

It was nice talking with you. Best of everything to you too.

Kay, as we can see, is a person who has latched on to a good career in a stable, but innovative, company. A few points stand out with regard to this interview.

First, did you catch what Kay said about punctuality? The SCRTD demands on-time performance from its personnel.

Second, note that Kay received all her electronics education and training in school. While she had exposure to electronics in a "hobby way" with her electronic art, she is basically a school-taught person. The message is that no one is born with a transistor in his or her hand. It's never too late to learn electronics.

Third, note Kay's concluding comment about enjoying electronics. She's saying that the interest and desire have got to be there. With that, almost any obstacle can be overcome.

Gideon Green: Office Equipment Field Service Representative

Gideon Green (Figure 4.5) is a 54-year-old field service representative/ area manager for Ricoh Business Systems. As such, he visits worksites in his area on a regular basis to do preventative maintenance according to the manufacturers' recommended schedules, and whenever emergencies arise. During these calls Gideon also advises customers on how to use equipment more efficiently and how to spot problems in their early stages. Furthermore, he listens to customers' complaints and answers questions, promoting customer satisfaction and good will. Gideon is concerned primarily with the maintenance and repair of high-end office fax machines. In addition to his technical duties, he supervises four to six field service technicians. For this interview, I accompanied Gideon on a service call to an optics company in North Hollywood, California.

Gideon, what is the problem here? Why are you making this call?

The customer is saying his fax machine is jamming all the time. This particular machine has two trays: one holds 500 pages, the other, 250 pages. The first thing I must do is try to get the machine to duplicate the complaint, that is, get it to jam.

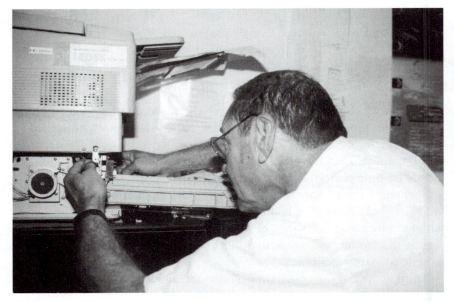

Figure 4.5
Gideon Green, field service technician.

What is the usual problem, why do such machines jam?

Sometimes a page is sticking up. That can cause a jam. The separation roller is supposed to pick up only one paper at a time. Sometimes the roller develops a flat area after being in frequent contact with paper. This roller looks alright, though.

So, what will you do now?

I am going to take a look at the pin clutch. That could be the problem.

Will you be able to know if the clutch is bad just by looking at it?

In many cases, yes.

What percentage of your repair work is mechanical?

Perhaps 60 percent mechanical and 30 percent electronic. This is a heavy-use machine, costing around $1,000. Often it is a mechanical problem.

Mechanical and electronic add up to 90 percent. What accounts for the remaining 10 percent?

User operator problems.

How do you deal with that?

Very carefully. Above all, you do not want to downgrade, or humiliate the customer. If you fix the machine, but the customer winds up hating you, what have you done?

What percentage of your job is customer relations?

About 50 percent. You must understand their needs. The machine is an important part of their business. Often, they are agitated, upset about the downtime. You must know how to diffuse the situation, restore calm, get the job done.

How important is it for you to have the same list of customers to visit?

Quite important. Ricoh is in transition on this. Eventually, we, the field service representatives, will have the same customers.

What do customers want from a good field service representative?

First, they want you to address the problem. But you must do it correctly, not rush through things. You must be thorough, complete. Ricoh has a philosophy, a motto: "We are here not to take a service call, but to prevent the next one." In other words, we fix what's wrong, but also eliminate what could go wrong.

Is that what makes Ricoh different?

Let's be honest. There is little difference between products from one company to another. So why does a customer go with one company over another? The answer is service. We must compete on service. That's what the customer wants, good service.

You're really Ricoh as far as the customer is concerned, aren't you?

Yes, I must never forget that. In many cases, the only dealing a customer will have with Ricoh is through me. It is an important responsibility.

Are you busy?

Yes. Machines do break down. These machines get a great deal of use.

So, in the end, what makes a good field service representative?

It's a combination of things. Mechanical and electronic ability is important, but it is not enough. The technician has to have communications skills, the ability to work with a customer. I do not want to hear complaints from a customer that a tech was impolite, screaming, etc. That is unacceptable.

What is your technical background?

I have been here 13 years. Actually, I got my initial training in the army, in Israel. I came to this country in 1976. I worked for Disney for a number of years. When I was laid off from Disney, I studied at Control Data Institute for a year. That's where I enhanced my electronics knowledge.

Tell me about your management duties.

I have four to six technicians working under me. It provides a chain of command. At Ricoh, there is a clear career path. It goes as follows: trainee, associate field service representative, senior field service representative, area manager, field manager, branch manager, and district manager.

So, field service representatives report to you?

Yes. If they have technical or other problems, they contact me first.

Tell me about your management style.

It is very important for me to know what is going on, where my techs are and what they are doing for workload distribution. It is important to keep in touch. My philosophy is this: I will give someone the benefit of the doubt. I assume he or she is a good person. Often it is just a matter of good guidance. A manager should be open, not hide in a cubical. Some of my reps will call just to talk, just to touch base. As a manager, you must understand why that is important to them.

Do technicians make good managers?

It depends. You must have the personality for it. All the management training in the world won't help if you don't have it in you, if it isn't a part of your makeup.

Looks like the fax machine has been operating well with the new clutch.

Yes. I think that does it. If it jams again, the customer knows how to reach me.

Gideon, as you can see, has made the transition from the strictly technical to management. He still fixes fax machines, gets his hands dirty (literally), and still makes field service calls. Yet, Gideon has the temperament, skills, and maturity to supervise and manage people as well. Ricoh Business Systems has taken advantage of those attributes to place him on a rewarding career path.

Eric Chavez: Owning Your Own Electronics Business

Eric Chavez is a 38-year-old, self-employed electronics technician who owns and operates his own electronics service business in a small town in southern California. After spending many years as a field and bench tech and after fixing stereos, TVs, VCRs, and other consumer electronics products in his garage for his neighbors and friends, Eric decided to go into business for himself. In the following interview, note what Eric (Figure 4.6) has to say about the business as well as the technical end of his operation. Is going into business for yourself something you might want to do someday? Reading this interview could help you to decide.

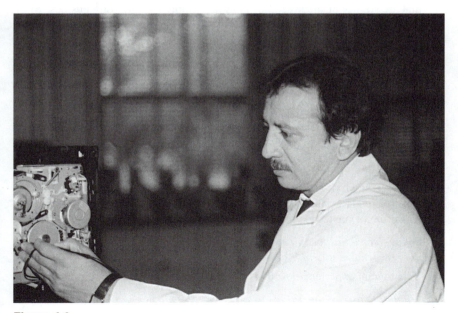

Figure 4.6
Eric Chavez, consumer electronics technician.

Eric, tell me about your business.

We are strictly a service operation. We service a variety of consumer electronic products, such as TVs, VCRs, stereos, fax machines, even Nintendo games and cordless telephones.

How long have you been in business?

Five years.

Who works for you?

I have three employees full-time. One works at the desk, taking phone calls, ordering parts, making up estimates. I have a full-time bench technician. And I have an outside man. He picks up TV sets, stereos, etc., and brings them to the shop. He may do minor repair in the field, the customer's home, but most products are brought back here for repair.

What are your responsibilities?

I run the business, of course. But I am also still a technician. Mainly I fix the TVs that my bench tech can't handle.

What percentage of your time is spent on business matters versus technical matters?

I would say 25 percent on business, 75 percent on technical.

What business skills are required to go into a business like this?

You have to possess both a business and a technical sense. There are those that are strictly businesspeople. They are simply looking for a business to buy, like a fast-food restaurant or donut shop. People like that have a lot of problems in this business.

Why?

They don't know how much time to spend fixing a product. They don't know how to come up with an accurate estimate. And that's the critical thing in this business, the estimate.

From a technical-knowledge standpoint, at what point is a technician ready to go into business for himself or herself?

Let me tell you my story. Through friends and neighbors, I started working in my garage fixing TVs, VCRs, and so on. I did it afternoons, evenings, and weekends. I said to myself: "What am I doing it this way for?" I knew it was time for me to open my own service business.

So an electronics technician who is working for someone else can begin doing work like this in his or her spare time?

Yes, to gain experience. And also begin to accumulate test equipment.

What are some of the pitfalls to watch out for when going into business for yourself?

I wasted a lot of time on Mickey-Mouse advertising because of lack of business experience.

What surprised you most about going into business for yourself?

All the work and the demands of customers—wanting it fixed now, etc. I was very surprised. There were times when I had to spend my holidays working.

Do you put in more than a 40-hour week?

Are you kidding? Often 60 hours or more.

But then you are your own boss. No one tells you what to do, right?

Yes, but there is another side to it. You have to be very self-disciplined. Sure, I don't have to go in to work. But if I don't, eventually I'll starve—or at least lose the business. Another thing: If you go fishing for 4 or 5 days, you better have a lot of confidence in your employees.

So what are the advantages of being in business for yourself?

You can make a lot of money if you run your business right.

How do you do that?

In this business the main thing is to get warranty work—get the products that are still under warranty.

Is that easy to do?

No, it is not. Becoming an authorized service center is not easy. You have to qualify. The manufacturers have strict rules. You must be in business for a long time. You must agree to attend their product-repair workshops on a regular basis. You must demonstrate technical expertise on a regular basis. You must also have sophisticated, and expensive, test equipment. The list goes on and on.

But it's worth the effort?

Yes, a lot of business then comes your way.

How is it having people work for you? How is it being the boss?

It's ok. Except when your technician comes to you and says, "I know more than you. Pay me more or I will go somewhere else."

What about the issue of customer relations?

It's very important. A tech has to know how to talk to the customer. For example, a customer brings in a TV set that is, let's say, 7 years old. You know the picture tube is bad and it will cost the customer $300 to have you replace it. How do you get the customer to approve that kind of estimate?

How do you?

Very, very carefully. You start by applying some customer psychology. "Do you have any sentimental attachment to the set? Was it given to you by your husband, brother, etc.?"

What, in your view, is the main reason why people fail in this business?

Honesty, or lack of it. Lack of honesty toward customers.

Plain old integrity?

Yes. For instance, a TV set comes in, no picture, no sound. I find it's a blown fuse or an open resistor. But I tell the customer I changed the flyback transformer, which, of course, costs more.

So, flat out lying?

Overcharging.

So then what happens?

It gets back to you. The customer takes it to another shop. They are honest about it. The customer spreads the word, writes a letter to the chamber of commerce. Eventually you're out of business.

Would a person be more likely to fail because of lack of business know-how or technical know-how?

I would say both. But since most people in electronics service will be stronger in the technical areas, they should be particularly aware of any weakness in business skills.

What do you think of partnerships?

I don't believe in them.

What's the problem?

Honesty. You have to really know your partner. Also, personality problems.

Does it cost a lot of money to go into a business like this?

A fair amount, maybe $50,000 to $75,000.

Do you have any final thoughts you would like to share?

Just tell your students that while it's a lot of work, it is also very rewarding. I'm glad I did it.

Thank you very much for your time, Eric, and may your business be a prosperous one.

My pleasure.

Well, what do you think? Is owning and operating your own business for you? Of course, you can't make such a major decision solely on what you've read here. And maybe that's the main point. It is obvious from what Eric has said that opening up your own electronics business is a serious undertaking, not to be done without a great deal of planning, knowledge, and understanding of the risks and rewards, and a fair amount of capital. Perhaps the best approach is to ease into it as Eric did, by working on friends' equipment in your spare time and in your garage. While you are doing that, it might be a good idea to take a few business courses at a local college. As Eric said, you're going to need both a business and technical sense to survive and prosper in your own business.

The five profiles presented here should have given you a feel for what it's really like out there, working as an electronics technician. If you liked what you read, then move on to Part II, where you'll discover how to become an electronics technician.

Summary

In Chapter 4, we met five electronics technicians. We talked with Ron Bagwell, an industrial electronics technician at the Los Angeles Times Olympic plant, a huge industrial enterprise. We met David Wells, where we discovered how important A+ certification is for electronics technicians. We met Kay Koopman, a hard-working communications tech at the Southern California Rapid Transit District. We saw how Gideon Green has combined both technical and management skills for Ricoh, a digital office automation company. And we found out from Eric Chavez what it is like to run your own electronics business.

Review Questions

1. At the Los Angeles Times Olympic plant, every aspect of newsprint production is controlled _____.

2. As an electronics technician, writing skills are important, because you often have to explain what you did to resolve a _____.

3. To work as an electronics technician, some employees make a _____, from one job category to another within the same company.

4. _____, or being on time is something employers often demand.

5. Gideon Green says that _____ percent of his work involves a knowledge of mechanics.

6. Since competing products are often alike in quality, price, and features, most companies wind up competing on _____.

7. According to Gideon, to be successful at management, you must have the _____ for it.

8. According to Eric, one of the critical skills for success in business is the ability to come up with an accurate _____.

9. According to Eric, one of the main reasons people fail in business is due to a lack of _____.

10. Opening your own electronics business is a serious _____.

Individual and Group Activities

1. Interview an electronics technician to obtain a profile of a typical working day. Take along a tape recorder and play back the interview to the class. Alternatively , prepare a 2- or 3-page report in dialogue form.

2. Interview someone who is in business for himself or herself—any business. Find out what motivates him or her, the advantages and disadvantages, the pitfalls to watch out for, and if the person would do it again. Give a 3- to 5-minute oral presentation of your results or summarize them in a 2- or 3-page written report.

3. Prepare a 5-minute video highlighting a technician on the job. Follow him or her around with the camera on a typical day. Show the video to the class and discuss the results.

Issues for Class Discussion

1. Discuss what questions it would be necessary to ask an electronics technician during an interview. Come up with a list of at least five such questions.

2. Discuss the Ron Bagwell profile. What stood out in your mind? What impressed you the most about Ron's job? The least?

3. Discuss the issue of making a lateral move within a company from an unskilled or semiskilled job to electronics. Is it wise to do this? What are the advantages and disadvantages? How does one find out about such openings?

4. Discuss David Wells' A+ certification. How did getting his certification affect his job prospects? (Note, Chapter 8 discusses IT certification in detail.)

5. Discuss the Kay Koopman profile. What stood out in your mind? What impressed you the most about Kay's job? The least?

6. Discuss women as electronics technicians. Are they any more or less likely to succeed than male technicians? Are employers more or less likely to hire women than men as electronics technicians?

7. Discuss the issue of honesty that Eric Chavez brought up. Is honesty always the best policy? What about the honesty of employees vis-à-vis the boss? What about honesty between partners?

PART

2

Becoming an Electronics Technician

In Part I we examined what it means to be an electronics technician. In Part II we explore the process of becoming an electronics technician.

In Chapter 5, "Taking the Big Step: Getting Educated in Electronics," we explore postsecondary educational institutions, both public and private, that offer electronics technician–level training. We investigate various curricula and programs, examine how to select an appropriate school, discuss the purpose and significance of school accreditation, and look into the question of financial aid.

With Chapter 6, "Succeeding as an Electronics Student: Making the Most of Your School Day," we begin by learning how to manage our most precious resource—time. We then discuss learning skills for electronics students: how to be more effective at reading, listening, mathematics, and circuit construction, testing, and troubleshooting. We conclude by examining key study strategies for grade-A learning.

Chapter 7, "Being an 'Electronics Activist': Electronics Beyond the Classroom," goes on to explain what can enhance and support formal electronics education. We probe project building, electronics organizations and clubs, amateur radio, and the issue of technician certification. We explore the value of attending trade shows and conventions, examine ways to keep up during summer vacation, and look at how the all-important information interview can be exploited with regard to the job search.

In Chapter 8, "Information Technology (IT) Certification: A New World of Credentialing," we examine why IT certifications are so popular and why they are relevant for electronics technicians. We take an in-depth look at A+ certification, while also exploring Network+ and i-Net+ IT certifications.

Finally, in Chapter 9, "Launching Your Career as an Electronics Technician: Finding, Getting, and Keeping Your First Position," we begin by tackling head-on the issue of "fresh out of school: little or no experience." We then talk about ways of finding, getting, and keeping a full-time position as an electronics technician.

Having completed both Part I and Part II, you'll not only know what it's like to be an electronics technician, you'll have the resources to become one.

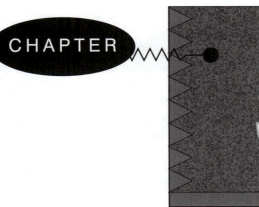

CHAPTER 5

Taking the Big Step:
Getting Educated in Electronics

Objectives

In this chapter you will learn:

- The difference between an associate and bachelor's degree.
- Why the associate degree has gained in value.
- How certificate programs are structured.
- About the role of 4-year and 2-year postsecondary institutions in your electronics education.
- About Tech-Prep, and how 2+2 and 2+2+2 works.
- How you'll be trained as an electronics technician.
- What an electronics core curriculum consists of.
- About job shadowing, internship, co-op education, and apprenticeship programs.
- How to find the best accredited postsecondary institution for you.
- What's available in grants, loans, scholarships, and work-study.

To be employed as an electronics technician—be it in biomedical service, industrial robotics, satellite communications, VCR repair, or computer maintenance—you need training and education in electronics. As a technician, you must be able to install, repair, test, maintain, and construct electronic products and services. While much of your knowledge and skill in doing so will come from practical, on-the-job experience, it is imperative that you first acquire a firm understanding of electronic component, circuit, and system operation, reinforced by experimentation and product construction, testing, and troubleshooting. To get this knowledge, you need to spend time in a controlled and organized environment, free of product-development and manufacturing restraints. The only place to do that is in school. Your formal education in electronics, more than any other single factor, will determine your success in getting that first job as an electronics technician.

In Part I, we saw what it is like to be an electronics technician. In Part II, we find out how to become that technician. Let's get started by seeing what the options are for an electronics education.

Options for an Electronics Education

To receive training and education in electronics at the technician level, a wide range of electronics programs are available across the country at a variety of institutions. There are certificate and diploma programs, two-year associate degree programs, four-year bachelor degree programs, and even home-study programs. (For an examination of IT certification, see Chapter 8.) In turn, various institutions—universities; colleges; community colleges; trade, vocational, and technical schools; and correspondence schools—are ready to deliver instruction in electronics. Let's begin by examining each program—its features, advantages, and disadvantages. Then we'll turn to the institutions that offer electronics training and education and examine their characteristics as well.

Since an **associate degree** has become the yardstick in the electronics industry by which technicians are measured and compensated, we'll explore this degree program first.

The associate degree had its genesis at the junior-college level many decades ago. It is a 2-year college degree, since junior colleges (also community colleges) typically offer a 2-year course of study. Traditionally, the associate in arts and associate in science degrees are awarded. Today, however, it is also common to see A.A.S. (associate in applied science) and A.S.E.T. (associate in science in electronics technology) degrees. Private as well as public postsecondary institutions offer the degree, although it is most closely identified with today's community colleges, of which there are over a thousand nationwide. It is a nationally, if not internationally, recognized degree.

The associate degree (Figure 5.1) requires a total of 60 to 64 semester units, the bulk of which are taken in your major if you are in a nonacademic, nontransfer program. In addition to electronics and related courses, you will take general-education courses (usually 12 to 15 units) in such areas as the natural sciences, social and behavioral sciences, humanities, language and rationality, and health.

But is getting an associate degree worth it? Up until recently, the associate degree was believed to be a nonessential entity and that completion of some technical courses were in and of themselves most important. Hence, many students passed up the degree. Later in their careers they regretted it. Since our society is credentials conscious, in retrospect they

Figure 5.1
Associate degree.
Source: Los Angeles Valley College and Ronald Reis.

learned that the degree requirements must be met. Recent trends show the associate degree is increasing in value—for those who earn the degree and for their employers.

Writing in *Community College Week,* Copper, Garmon, and Kubala said: "Employers understand that in the occupational-technical area, associate degree programs are complete—not half of something—and are designed to develop competencies in individuals that are immediately useful on the job." In a *Miami Herald* newspaper article, it was made clear that the wave of the future is the associate degree. "Employment of technicians and related support occupations is projected to grow by 22 percent, adding 1.1 million jobs by 2008," it said. The implication is clear: If technician jobs are increasing, so will the need for the associate degree.

However, if you are looking for very directed and specific training in electronics, a **certificate** or diploma program may be more to your liking. Such programs, whether offered by community colleges, trade, vocational, or technical schools, or correspondence schools, recognize a student's satisfactory completion of an organized program of vocational study. Courses required to earn the certificate or diploma are almost entirely in the major field of study. Certificate programs in electronics can take from a few months to 2 or 3 years to complete, depending upon the course of study and type of institution offering the program.

Certificate programs are quite useful for recognition of specific training in a focused field of study, from CAD expertise to quality assurance testing. Indeed, there is often no other way to acquire such explicit job-related training.

Swinging the other way in terms of length of study, we have the **bachelor's degree,** representing the attainment of a traditional four-year-college education. As we saw in an earlier chapter, those desiring to do technician-type work but also have the advantages of a college degree might want to pursue a major in engineering technology, applied technology, or industrial studies at the college or university level. Such programs require you to take a full complement of general-education courses (40 to 45 units), 15 to 20 units of electives or a designated minor, and the remaining courses (60 or so units) in your major. With a 120- to 128-unit requirement, you can see why it takes at least 4 years to complete a bachelor's degree.

Another type of electronics program, which is often underrated, is **home study.** Courses given in correspondence programs can range from the inadequate and frivolous to well-established and recognized programs. With correspondence courses, you are usually sent a package of materials containing electronic components (and, in some cases, equipment), a textbook, workbook, notebook of lessons, audio and video tapes, study guides, and even the phone number of your professor, whom you are encouraged to contact when questions or problems arise. Home-study courses find application at both ends of the electronics training spectrum—that is, at the beginning, when you might want simply to explore the field, and after you have been on the job for some time and want to update your knowledge in a specific skill area.

As you might expect, there are a number of institutions ready and willing to offer training and education in electronics. The most relevant for the would-be electronics technician are community colleges and trade, vocational, and technical schools. We will examine these institutions more closely later when we deal with the school search. For now, let's briefly discuss all institutions, with particular reference to the type of electronics programs they offer. In doing so, we will refer to Table 5.1, which cross-references types of institutions and programs. We must keep in mind, however, that what is discussed here is a generalization. There are exceptions in almost every case. For example, the place to earn a bachelor of science degree in electrical engineering (B.S.E.E.) is ordinarily at a university. Yet although the California State University at Los Angeles (CSULA) offers such a degree, it also awards a B.S. in engineering

TABLE 5.1
Electronics education programs and institutions.

	Institution				
Program	**University**	**Four-year College**	**Community College**	**Trade, Vocational, Technical School**	**Correspondence School**
Certificate	No	No	Yes	Yes	Yes
Associate degree (A.A., A.S., A.A.S., A.S.E.T.)	No	No	Yes	Yes	Some
Bachelor's degree (B.A., B.S.)	Yes	Yes	No	Some	No
Home-study diploma	No	No	No	No	Yes

technology and a B.S. in industrial studies, and issues a certificate of completion in electronics technology.

Beginning with the **university,** we note that it offers B.A. and B.S. degrees, master's degrees (M.A., M.S.) and, in most cases, doctorate degrees (Ph.D.). Entrance requirements tend to be stiff, as are costs, especially at private universities. In all but a few cases, the university is set up to produce electrical engineers, not electronics technicians.

Four-year colleges will, in most cases, offer a B.S.E.E. Unlike many universities, however, they often have engineering technology and industrial studies–type programs. Entrance requirements are, on the whole, less stringent than those of the university; costs, at least for public colleges, may be slightly lower than at the university.

Community colleges, also known as junior colleges, offer associate degrees and, in most cases, certificates of completion. Entrance requirements are minimal; a high-school diploma or being 18 years of age or over and having the ability to profit from the experience is usually all that is necessary. In many states, however, a local residence requirement is in effect. Average annual tuition and fees range from $100 to more than $2000, with an average of $1518. The community college is structured to train and educate future electronics technicians. For most students seeking to become technicians, the local community or junior college is a viable option.

Nonetheless, **trade, vocational,** and **technical schools**—for the most part private, although some are public—offer practical alternatives to the community college for training electronics technicians. Proprietary (private) schools, such as the DeVry Institutes of Technology and the ITT Institutes of Technology, offer associate degrees and certificates (or diplomas). Such schools are designed specifically to train individuals for careers as electronics technicians. Entrance requirements vary widely and are usually based on a combination of factors, such as high-school grades, math background, entrance-exam scores, and teacher recommendations. Costs at private schools tend to be higher than those at public institutions, although loans and grants are often available.

Correspondence schools, as one would expect, vary widely in scope and quality. The better ones offer diplomas, certificates, and even associate degrees. Entrance requirements are usually nonexistent—you must simply be able to benefit from the course work, a decision left entirely up to you. Costs also vary, but as a rule they tend to be quite high. It is worth keeping in mind, however, that courses often come with electronic supplies and equipment (meters, oscilloscopes, etc.) as part of the cost.

For those just beginning their study of electronics, universities, 4-year colleges, community colleges, trade, vocational, and technical schools, and correspondence schools all offer training and education. Furthermore, throughout the country, a combined secondary and postsecondary program based on a formal articulation agreement providing students with a nonduplicative sequence of progressive achievement leading to technical competencies has emerged. Known as Tech-Prep, it is widely praised. Let's see why.

Tech-Prep: Linkage That Works

The son of immigrant parents, one of seven siblings, and living in an inner-city housing project, not much was expected of Jose. Even if he graduated from high school, a 50-50 chance at best, he'd probably wind up going from one no-brainer job to another, earning minimum wage, maybe. Assuming he could ignore or avoid the degradation and gang life around him, there still was no way, according to the conventional wisdom, that Jose was going to achieve the self-respect and income necessary to raise a family and capture a snippet of the American dream.

Yet, Jose was not a loser. As he entered high school, this smart young man developed a keen interest in electronics, particularly computer repair. All he needed was a program to capitalize and channel his zeal.

Had Jose gone to a traditional high school, the kind you may have attended, chances are any enthusiasm and motivation he developed would soon have been dissipated. Back then, high school divided students into two groups: college prep and no prep. Or, more bluntly, winners and losers.

Fortunately for Jose, the high school he attended had a new attitude, a new way of doing things. Not wanting to attend a 4-year college, at least not at first, Jose wished to enter the workforce as soon as possible as an electronics technician. At Jose's high school, non-college bound students like him, 75 percent of the student body, had an alternative to college prep. Called Tech-Prep, it is the antidote to no prep. The program put Jose on the path to occupational success.

So, what exactly is Tech-Prep, and how can it benefit you? Tech-Prep prepares a student for a highly skilled technical occupation that allows either direct entry into the workplace as a qualified technician after 4 years, or continuation with further education leading to a bachelor's degree after 6 years.

In the 4-year program, known as 2+2, a student begins Tech-Prep in the last 2 years of high school (11th grade) and continues with 2 years at a postsecondary institution, usually a community college. In the end, he or she earns a certificate or associate degree. Hopefully, the student now possesses the knowledge and skills necessary to work in a technologically advanced society.

In the 6-year program, known as 2+2+2, a student continues on to a 4-year college as a junior after earning an associate degree from a community college. He or she can receive a bachelor's degree in fields such as electronics or industrial technology.

Remember, with Tech-Prep, you don't have to start in high school. If you're already at a community college and are thinking about transferring, contact a counselor right away. See if Tech-Prep is for you.

The Electronics Curriculum: How You'll Be Trained and Educated in Electronics

It's time now to investigate the electronics curriculum itself. In doing so, we begin by examining a typical 2-year associate degree program as offered at a public community college or a 2-year trade, vocational, or technical school. We then analyze course content and discuss the issue of course transferability. Finally, we will go into support programs that can enhance any electronics curriculum. We refer here to job shadowing, internship, cooperative education (co-op), and apprenticeship programs.

Coming up with a typical 2-year electronics curriculum is not easy: There is a great diversity of programs across the land. Nonetheless, it is possible to present a curriculum that contains most of what you are likely to find at any 2-year postsecondary institution. Such a curriculum is diagrammed in Figure 5.2. Two factors should be kept in mind. First, if the school you plan to attend or are now attending does not offer everything you see here, do not be dismayed. Your school may be specializing in one or two areas, or, more likely, it may simply categorize its courses differently. Second, specific subject matter crosses or penetrates many courses. For example, you are likely to encounter the study of microprocessors in any advanced electronics course you take.

Most 2-year programs have a group of **core courses** and a group of **advanced courses**. The former are taken in the first year; the advanced, in the second year. The core courses are required of everyone in the major. They are designed to give you a firm foundation in basic electricity and electronics. Such courses are taught in a lecture and laboratory environment.

With the majority of electronics technician programs, you have a choice of advanced courses to take. In other words, in your second year

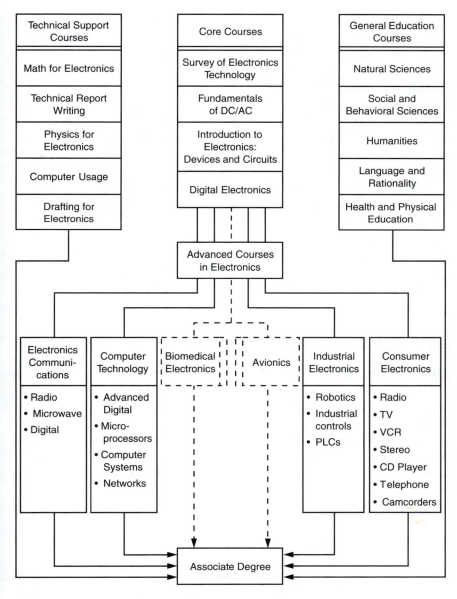

Figure 5.2
An electronics curriculum.

you begin to specialize: radio communication, industrial electronics, computer technology, consumer electronics, or even such specialties as biomedical electronics, music technology, or avionics.

Paralleling electronics course work, both core and advanced, will be **technical support** and **general-education courses.** The former consist of such offerings as math for electronics (usually algebra-based), technical report writing, physics for electronics, computer usage, and electronics drafting. In completing your general-education graduation requirements, you will take courses in the natural, social, and behavioral sciences, humanities, language arts, and health and physical education.

In the end, all these areas of course work are combined, as in Figure 5.2, and you are awarded the associate degree. While it would serve little purpose here to detail the contents of every course listed in Figure 5.2, it might, nevertheless, be helpful to clarify the subject matter of the core-course offerings, since they are almost universally required of the electronics technician major. A description and comment regarding each core course follows.

Survey of Electronics Technology

This course is not offered by all institutions, some preferring to jump directly into dc and ac. Where it is given, it tends to be either a true survey of the subject or a "stretched-out" version of the dc/ac course. As the former, it will delve into the electronics industry, the people in it, career opportunities, and the role of electronics in our lives. A laboratory component is not always present. If the course is more a gradual introduction to dc and ac, it can help the hesitant student ease into the major. In this latter version, lab time is almost always provided.

Fundamentals of Electricity: DC/AC

This may be one course in one semester or two courses in a two-semester sequence. It is the foundation course in the electronics technician curriculum. In the dc portion, you study dc circuits, Ohm's law, energy sources, magnetism, capacitance, inductance, Kirchhoff's laws, Thevenin's theorems, and bridge circuits. In the ac section, alternating current principles, LCR circuits, resonance, transformers, filters, and band-pass circuits are the key topics covered. Both elements have a strong laboratory component, where you investigate electrical components and circuits and learn

Figure 5.3
Electronics school laboratory.

how to use meters, power supplies, signal generators, and oscilloscopes.
See Figure 5.3.

Introduction to Electronics: Devices and Circuits

This is the first true electronics, as opposed to electricity, course. Here you
will scrutinize diodes, transistors, and linear integrated circuits. You will
also explore the theory and operation of power supplies, amplifiers, and
oscillators. Emphasis in the laboratory is on circuit-construction techniques and troubleshooting.

Digital Electronics

Digital electronics used to be an advanced course in most electronics programs and still is in some. But today the trend is to bring digital electronics into the first year as a core course to be required by all those in the
major. Digital electronics is such an important part of every aspect of electronics that it is paramount for all would-be electronics technicians to be

exposed to its circuit elements: logic gates, combinational logic, sequential logic, and semiconductor memories. Lab time is spent building, designing, and troubleshooting digital circuits.

The question often arises as to which, if any, of the electronics courses in a standard 2-year electronics technician curriculum (as opposed to a lower-division engineering program) are transferable to a 4-year college or university. It depends on two factors. One is the type of program to which you are transferring: a bachelor of science in electrical engineering, engineering technology, or industrial studies. The other is the "articulation" agreement your 2-year school and the specific 4-year college or university have worked out concerning what courses are accepted and how much credit is given for them.

Generally, you will have the hardest time transferring courses to a B.S.E.E. program and the easiest time bringing them to an industrial studies curriculum. The reason is simple. The electronics courses taught in the B.S.E.E. program are, almost without exception, *calculus-based*. The textbooks are written and the lectures given with calculus in mind. The courses are taught with the assumption that students have had advanced mathematics. If the course you have taken at a 2-year institution does not have a calculus prerequisite, and most offered in an electronics technician program do not, then it is unlikely the course will transfer. This isn't necessarily the case when transferring to an engineering technology or industrial studies program. Electronics courses for these latter two degrees (particularly the industrial studies or applied technology programs) rarely have as stringent a math and science requirement as the B.S.E.E. program.

How do you find out for sure what is or is not transferable? The best approach is to contact a responsible person in the department of the school to which you wish to transfer and discuss the matter directly. Send him or her a description of the courses you are or will be taking at your 2-year school. Then ask for validation of their transferability. Sometimes it becomes a matter of negotiation between you and the department as to exactly what can be accepted. Do this as early in your studies as possible so that you know up front what is transferable and what is not.

In addition to offering courses in electronics and related subjects, many 2-year postsecondary institutions seek to provide students with a link between school and work in order to enhance exposure to and experience with the career of an electronics technician. In the concluding portion of this section, let's examine how *job shadowing, internship, co-op education,* and *apprenticeship* programs operate to fulfill these goals.

You've heard of Groundhog Day, but what about Groundhog Job Shadow Day? This is when you see what it's like to work in an organization that interests you by spending half a day at a leading local company or organization, "shadowing" someone who does that line of work. While one can job shadow any time of year, this nationally sponsored day draws over a million students.

Basically, the idea is to:

* Demonstrate the connection between academics and careers, excite students to learn by making their classroom work more relevant.
* Build community partnerships between schools and businesses that enhance the educational experience of all students.
* Introduce students to the requirements of professions and industries to help them prepare to join the workforce for the twenty-first century.
* Encourage an ongoing relationship between students and employers.

When students visit organizations, some of the ideas discussed are:

1. A little history of the career or field
2. Roles and responsibilities
3. Personal attributes
4. Career ladder
5. Education requirements and opportunities
6. Technology
7. Related jobs and careers
8. Learning more about this career
9. Your experience
10. Money questions

Sound interesting? Contact your local community college or go to the Groundhog Job Shadow Day website, at www.jobshadow.org/, for more information.

Typically, an **internship** program involves placing a student in a *nonpaying* work environment for a few hours a week. The twofold purpose is to (1) give the student exposure to the field and (2) provide a chance for the company and the student to "check each other out" with regard to possible employment later on. Both of these goals were advanced when Michael McCray, a 19-year-old, second-semester student at Los Angeles

Trade Technical Community College linked up with Target Communications Systems in El Monte, California, through a 100-hour, 10-hour-a-week intern program. According to a satisfied Michael, "When I arrived at Target, I was a bit nervous and didn't know what to expect. But right away they put me to work, sort of on a buddy system, with Sam, an experienced installer. I went out with him and helped install commercial radios in cars and vans. At first I just watched and handed across the tools. But Sam taught me everything he could, and by the time my 100 hours were up, I could do basic radio installation. I guess they liked what I did because they offered me a full-time summer job working as an entry-level installer for $9.50 an hour. I took it, and it has worked out great."

For an internship program to succeed, four elements must be present:

1. The student should do meaningful work; he or she should not just "hang around" and look over a technician's shoulder.
2. The student should not be stuck in a narrowly defined, repetitive task where exposure to the occupation and the company is circumscribed.
3. The employer must not view the student as free labor to be exploited.
4. The job supervisor should take an active interest in the student's learning on the job.

If these elements are met, then an internship program can add considerably to a student's electronics education.

Cooperative (co-op) education takes internship one step further—to a paying job. Furthermore, in a co-op program you're likely to be placed in a clearly defined job category, where you will work and earn just the same as everyone else in the company. In co-op programs, the school takes the leading role and the employer is enlisted to "cooperate." The intent is to create a close connection in the student's mind between the job and the classroom. In 1996, 429 two-year colleges placed 81,000 students (77 percent vocational) in co-op programs.

Typical of such programs is one run by the Federal Aviation Administration (FAA). As you may know, the FAA is responsible for operating and maintaining the world's largest and most advanced air traffic control and navigation system. According to the FAA *(For Students Only: Cooperative Education)*:

> The FAA's Cooperative Education (co-op) program offers the best of two worlds—scholastic knowledge and practical experience. Our program offers on-the-job experience with pay while you attend

school. It's your opportunity to preview a career in a realistic work setting while you check out the FAA as an employer. Many co-op students are offered full-time, permanent career positions with the FAA upon successful completion of their educational and work requirements. You won't just spin your wheels in this co-op program; you'll get a head start on your career.

When hired into the co-op program by the FAA, students work 1000 hours in the second year of school. They are employed as radar, automation, communications, and navigation electronics technicians.

Apprenticeship programs, though not widespread in electronics technician training or education, are, nonetheless, available through some schools, most notably those in the Midwest and on the East Coast. Apprenticeship, according to the IBEW (International Brotherhood of Electrical Workers, "The Electrical Worker's Story") is

> . . . a system of learning which provides planned, day-by-day, on-the-job experience under the qualified supervision of a skilled craftsperson, along with related instruction that will teach you the technical aspects of the trade. It is a training program that includes actual paid employment in occupations where knowledge of specific related skills and technology is essential.

At Walt Disney Imagineering, where they create three-dimensional animated Disney characters, an apprenticeship program is operated in conjunction with local community colleges. After a student receives an associate degree in electronics, he or she is hired and placed in an apprenticeship for 2 years. The new employee is rotated through such areas as model making, woodcraft, electronics construction, hydraulics, plastics, electrical, and molding to give him or her a feel for the different skills necessary to produce the animated models used at the company's theme parks, such as Disneyland and Walt Disney World. After the apprenticeship program, the employee settles into the job of skilled electronics assembler and troubleshooter.

Apprenticeship, if available, can provide an excellent school-job link. Check to see if your school has such a program.

The School Search: Finding What's Best for You

As a prospective electronics technician student, you're in an enviable position. Unlike some of your friends who might be desperately seeking admission to a "prestigious" college or university, in your case the shoe is on the other foot, so to speak. You are not seeking to be chosen; you are doing the choosing. If you wish to attend a community college or a trade, vocational, or technical school, rest easy, you are going to get in, you will be accepted. The question becomes not whether they will choose *you*, but rather which one of *them* you will pick.

While that's good news, the fact that you're in the driver's seat means that you have an obligation to make the effort to select wisely and carefully. Once you let it be known what you are seeking in the way of an education, you will be besieged by various institutions, private as well as public, with all sorts of claims as to their ability to give you the best and most appropriate electronics education.

As we have said earlier, if you are a typical would-be electronics student striving for a career as an electronics technician, your choice of schools will boil down to a community college or to a private trade, vocational, or technical school. In making your selection, you will be governed by many factors, two of which, *time* and *money,* are paramount— that is, how much time do you have to spend on getting an education and how much will it cost you?

Generally, a community-college associate degree program will, at a minimum, take 2 years, full time, with summers off. The cost will average around $1500 a year for tuition and books. Private trade, vocational, and technical schools, on the other hand, can shorten that time to around 15 to 18 months (with no summer vacation). The cost, however, is considerably higher than at a public community college.

Regardless of which way you go, community college or private trade, vocational, or technical school (even correspondence school, 4-year college, or university), there are three key factors to address in your selection process. One, you must proceed with a three-stage **school investigation** that involves research, a campus visit, and a personnel interview. Two, you must determine if the school is **accredited** and by what agency. And, three, you should find out about **financial aid:** grants, loans, scholarships, and work-study programs. Let's take each factor in turn.

You will usually begin your school investigation by conducting research from your home base. Start by obtaining the school's most important explanatory document—its catalog. A list of community colleges with their addresses can be found in most collegiate dictionaries. For a similar list of private accredited trade, vocational, and technical schools, write to the Career Colleges Association (CCA), formally the National Association of Trade and Technical Schools (NATTS), 106 Wisconsin Avenue NW, Washington, DC 20002. When you receive the catalog, devour its contents. You'll learn more about an educational institution from this one document than from all others combined. But one word of caution: Do not assume that all courses listed in the catalog are indeed offered. To know what is actually being taught, ask for a current schedule of classes as well.

Beyond the school catalog, you can obtain valuable information about any prospective postsecondary institution from a high-school guidance counselor or from friends and alumni. The latter two are particularly important. If you can talk with persons who are attending or have attended the school, don't pass up the opportunity. Listen carefully to what they have to say. Try to determine "where they are coming from," what their likes and dislikes are, what prejudices, if any, they may have. Current students and alumni can give you valuable inside information that is difficult to obtain anywhere else.

After you have done all the research possible from your home base, it is time to visit the school's campus (Figure 5.4). When you do so, remember to write ahead for an appointment, visit the school when it is in session, and bring along your parents or a friend to bounce observations off of as you look around. In addition, here is a short checklist of what to observe and do while on campus:

Figure 5.4
School campus.
Source: Los Angeles Trade–Technical College.

1. Note the surrounding atmosphere. Is the campus peaceful or overly rushed?

2. Visit the library or learning resource center. Is it a quiet place with a wide variety of learning aids at the students' disposal?

3. Check out the electronics department's lecture and laboratory facilities.

4. If possible, sit in on one or more classes. Observe both the instructor's approach and the students' responses. Note, particularly, how the teacher handles student questions. Does the instructor take the time to zero in on the students' specific concerns?

5. Talk with students. Try to get a feel for what they think about the school. Look beyond what they say to their expressions of enthusiasm (or lack of it).

6. Pay attention to what's going on and ask lots of questions. Be assertive.

7. Find out about extracurricular programs. Are there any you may want to become involved in?

The third stage in your school investigation involves an interview with school personnel. For you, a prospective electronics technician student, that will probably mean the electronics department head or an instructor in the department. Remember, unlike your friend seeking admission to the prestigious college or university who is now "sweating" *the* campus interview, you're approaching the interview from the opposite direction. You are the interviewer, not the interviewee. Nonetheless, you still have to be prepared with thoughtful, relevant questions. The following is a list of such questions prepared by CCA. Needless to say, these questions apply to public as well as private institutions. Some of the questions may have already been answered during your earlier investigations. But if not, now is the time to ask the questions and get answers.

How long does it take to complete the program?

What are the choices for attending class? Are classes offered during the day and at night?

What skills will be taught?

How much of the technical training is hands-on, and how much is lecture?

How many students are in a class?

What equipment is used?

Is there an opportunity to get work experience?

Is tutoring available?

Can a student repeat a part of the course if needed?

Is a high-school diploma or GED required?

Is there an admission test, and how are the scores used?

Is there another interview before acceptance?

What information does the school need before acceptance?

What is included in the tuition?

Is the registration fee separate?

What are the estimated expenses for books and materials?

Are tools provided, or must they be paid for separately?

What has been the school's placement rate for the last year or two?

What companies have hired graduates?

What does the placement office do to assist in locating part-time jobs for students while they are in school?

Add to this list any additional questions you want answered. For example, while 95 percent of all community colleges are connected to the Internet, will you, as a student, have access? Remember, don't be timid. One way or another, get your questions answered. Keep in mind that it is your career at stake.

Next, let's turn to the all-important matter of *accreditation*. Accreditation means that a school has met national standards of educational performance established by an impartial nongovernment agency. Note the terms *national standards* and *nongovernment agency*. The former means that the accreditation agency operates nationwide, unlike licensing boards, which are state-run. Nongovernment means that the agency is private, although it may be approved by the government. Indeed, for you to be eligible for state and federal grants, loans, scholarships, and work-study programs, the agency responsible for accrediting your school must be recognized by the U.S. Department of Education.

There are many such accrediting agencies. Most are members of an umbrella organization called the Council on Postsecondary Accreditation (COPA), a nonprofit, private organization devoted to coordinating and improving accreditation. With regard to community colleges and vocational, trade, and technical schools offering programs in electronics, two accreditation bodies are of particular importance. These are the

Accreditation Board for Engineering and Technology (ABET) and the Accrediting Commission of the Association of Independent Colleges and Schools. Look to see if your school of choice has either of these accreditations. If it does not, however, it certainly doesn't mean that the school is unacceptable. Other perfectly credible accreditation agencies may be involved. For example, the school at which I teach, Los Angeles Valley College, is accredited by the Western Association of Schools and Colleges.

Finally, as we close Chapter 5, let's look at the very critical issue of *financial aid*. The cost of a postsecondary education is high and is growing at a rate above the inflation rate each year. The cost includes tuition and fees, room and board, travel, books and supplies, and personal and miscellaneous expenses. Many students, whether they attend a local community college or an Ivy League university, are doing so with some financial aid from federal, state, local, and private sources. In fact, about 35 percent of all community colleges students receive some kind of financial aid. By definition *(A Helping Hand: Your Guide to Financial Aid Programs and Services in the Los Angeles Community Colleges)*:

> Financial aid is monies made available by federal and state governments and private sources in the form of grants, loans, scholarships, and employment. These monies are available to make it possible for students to continue their education beyond high school even if they and their family cannot meet the full costs of the postsecondary school they choose to attend. The basis for such programs is the belief that parents have the primary responsibility of assisting their dependents in meeting educational costs and that financial aid is available only to fill the gap between a family's contribution and the student's yearly academic expenses.

We cannot hope, here, to detail all the available financial aid. What we will do is call attention to the four key types of financial assistance: grants, loans, scholarships, and work-study. If any of them seem appropriate to your needs, seek further information from a school guidance counselor or the agency in question.

Grants

A **grant** is the most sought-after form of financial aid because it is *aid that you do not have to pay back*—it is not a loan, it is out- right assistance. The Pell Grant is typical. It is a federally funded program that is available to undergraduates who can demonstrate financial need. In the Los Angeles Community College District, Pell Grants provide from $200 to $2400 per academic year.

Loans

A loan is financial aid that *does have to be paid back*. Typical of federally sponsored loans is the Perkins Loan. For students in the Los Angeles Community College District, up to $4500 a year may be borrowed. Repayment status begins 9 months after the borrower graduates, withdraws, or ceases to be at least a half-time student. Five percent interest is charged on the unpaid balance of the loan principal.

Scholarships

Scholarships are available for a number of reasons: good grades, financial need, athletic prowess, course of study, ethnic and racial identity and so on. Many scholarships go untapped. According to a recent study, there are more than 200,000 academic and vocational scholarships—worth about $15 billion. But more than $6.6 billion goes unused every year. If you are Greek, American Indian, Polish, Syrian, Japanese, Latino, or African-American, or if you or your parents are members of a professional or social organization such as the Kiwanis Club or the American Legion, you may qualify for one of the thousands of scholarships that go begging. It's worth looking into.

A relatively new national scholarship, known as the HOPE Scholarship Credit, should be of particular interest to community college students. Through it, taxpayers may be eligible to claim a nonrefundable HOPE Scholarship Credit against their federal income taxes. The HOPE Scholarship Credit may be claimed for qualified tuition and related expenses of *each* student in the taxpayer's family who is enrolled at least half-time in one of the first two years of postsecondary education and who is enrolled in a program leading to a degree, certificate, or other recognized educational credential. There are, of course, eligibility requirements. But it's worth checking out.

Work-Study

Work-study provides a job, usually on campus, while you go to school. The federally sponsored College Work-Study Program (CWS) is one example. It enables students to earn part of their financial aid awards through part-time employment. To be eligible, a student must be a U.S. citizen or permanent resident, must be enrolled in the appropriate number of units, and must maintain good academic standing while employed under the program. Hourly wages may vary with the type of work, which is available to students qualifying for financial aid during the summer.

Whatever your financial status, you may want to look into the various forms of financial aid discussed here. Take all the help you can get while getting your education in electronics.

When it comes to scholarships, grants, and even loans, however, you need to be cautious. Fraud can be a problem. Here, from the National Fraud Information Center, are six signs that a scholarship may be suspect:

1. **"The Scholarship is Guaranteed or Your Money Back."** No one can guarantee that they'll get you a grant or scholarship. Refund guarantees often have conditions or strings attached. Get refund policies in writing—before you pay.

2. **"You Can't Get This Information Anywhere Else."** There are many free lists of scholarships. Check with your school or library before you decide to pay someone to do the work for you.

3. **"May I Have Your Credit Card or Bank Account Number to Hold This Scholarship?"** Don't give out your credit card or bank account number on the phone without getting information in writing first. It may be a set-up for an unauthorized withdrawal.

4. **"We'll Do all the Work."** Don't be fooled. There's no way around it. You must apply for scholarships or grants yourself.

5. **"The Scholarship Will Cost Some Money."** Don't pay anyone who claims to be "holding" a scholarship or grant for you. Free money shouldn't cost a thing.

6. **"You've Been Selected by a National Foundation to Receive a Scholarship, or You're a Finalist in a Contest You Never Entered."** Before you send money to apply for a scholarship, check it out. Make sure the foundation or program is legitimate.

Summary

In Chapter 5, we began by looking at options for an electronics education, with emphasis on the associate degree. We surveyed 2- and 4-year college programs, as well as correspondence schools and home study. We examined the role of Tech-Prep in 2+2 and 2+2+2 articulation.

Next, we explored the basic electronics curriculum, in particular core courses. We talked about transfer from a 2- to a 4-year college and investigated job shadowing, interning, co-op education, and apprenticeships. We reviewed factors to consider when selecting a postsecondary institution for study. Finally, we delved into financial aid: grants, loans, scholarships, and work-study.

Review Questions

1. The _____ _____ has become the yardstick in the electronics industry by which technicians are measured and compensated.

2. The _____ _____ represents the attainment of a traditional four-year college education.

3. _____ _____, also known as junior colleges, offer associate degrees and, in most cases, certificates of completion.

4. _____-_____ prepares a student for a highly skilled technical occupation that allows either direct entry into the workplace as a qualified technician after 4 years, or a continuation with further education leading to a baccalaureate degree after 6 years.

5. Most 2-year electronics programs have a group of _____ courses and a group of _____ courses.

6. Typically, an _____ program involves placing a student in a nonpaying work environment for a few hours a week.

7. _____ (co-op) education takes internship one step further—to a paying job.

8. _____ is a system of learning which provides planned, day-by-day, on-the-job experience under the qualified supervision of a skilled craftsperson.

9. _____ means that a school has met national standards of educational performance established by an impartial nongovernment agency.

10. Financial aid involves _____, _____, _____, and _____-_____.

Individual and Group Activities

1. Pick a local 4-year college or university that offers both a bachelor of science in electrical engineering and a bachelor of science in engineering technology degree. Visit both departments to find out what distinguishes the two degree programs. Does one have more status than the other? Do the B.S.E.E people resent the presence of the bachelor of engineering technology program? Is one program "tougher" than the other? Present your results to the class in an oral report.

2. If you are currently attending a community college or trade, vocational, or technical school, find out if it is accredited and, if so, by what agency. When will the school be up for reaccreditation review? Is the school having problems with its accreditation?

3. Write to the National Home Study Council, 1601 18th Street NW, Washington, DC 20009 for a start in finding out about home-study programs. Interview individuals who have gone the home-study route. What are your views on home study? Summarize your findings in a 2- to 3-page report.

4. If you are attending a postsecondary institution, does it offer job shadowing, internships, cooperative education programs, or apprenticeships? If so, find out about them. What are the entry requirements? What duties and responsibilities do the student, industry, and the school have? What are the advantages and disadvantages of the programs? Present your findings to the class in an oral report.

5. If you are now working toward an associate degree, but plan to transfer to a 4-year college or university, determine what courses will be transferable. Prepare a list of those that will be accepted by your college or university of choice. What courses, if any, were rejected? Summarize your findings in a 2- to 3-page report.

6. If you are not enrolled in a postsecondary institution, visit one and record your observations. Prepare a list ahead of time of things to observe and do while on campus. Summarize your findings in a 2- or 3-page report.

Issues for Class Discussion

1. How is one to distinguish the good from the bad proprietary schools? That is, how do you separate the well-established, nationally recognized institutions from the local for-profit schools seemingly set up by operators to secure government loans for students who, for the most part, neither complete the courses nor are given rebates on their fees?

2. Discuss the differences between training and education. When would one be more appropriate than the other? Why are vocational schools identified more with the former than the latter?

3. Discuss how you feel about general-education course requirements. Do you resent having to take such courses? How might they contribute to a well-rounded technician? Is such a technician needed?

4. Compare and contrast an associate degree and a certificate of completion. What are the pros and cons of each?

Succeeding as an Electronics Student: Making the Most of Your School Day

Objectives

In this chapter you will learn:

- How to manage your time more wisely for optimum study.
- How to read a textbook the correct way.
- How to use your classroom instructor by asking the right questions.
- How mathematics can solve problems, the easy way.
- How reading a schematic is as simple as knowing a few symbols and how they're connected together.
- About circuit test and troubleshooting techniques.
- What study strategies will work best in learning electronics.
- How to master the test-taking game.

You have read Part I, "Being an Electronics Technician," you have talked to family and friends, you have located a good community college or technical school close to home, and, perhaps, you have even interviewed a technician or two. As a result, you have made up your mind—you want to be an electronics technician. The time has finally arrived to begin your studies in electronics. But as you do so, you become a bit apprehensive, even a bit nervous. You have some doubts. You might ask yourself, "Is this really for me? Can I actually do it?"

After all, maybe you have been out of school for awhile. Can you honestly buckle down and study? You didn't take electronics in high school; you actually know very little about it. You haven't built anything in electronics. You were never an electronics hobbyist or a kit builder. Sure, you once helped a friend install a car stereo, but all you really did was hand over the tools—when you could identify the ones your friend wanted, that is. You've never repaired anything electronic. Heck, you don't even know how to solder.

If any of this sounds familiar, you must be asking: "Have I truly made the right choice?" Furthermore, what about the actual study of electronics: the math, the endless list of confusing and unfamiliar technical terms, the "heavy" technical reading? Electronics can seem so complicated. "How can anyone ever understand even a tiny fraction of it?" you must be thinking. To be sure, it's all enough to make anyone uneasy.

However, don't be dismayed, don't despair, don't panic—you can learn electronics. Hundreds of thousands of people, with widely varied backgrounds, do it every year. So can you.

In this chapter we try to relieve any anxiety, apprehension, self-doubt, or jitters you may have. We show you how to find the time to study, how to develop critical study skills applicable to electronics, and how to do grade-A work in your chosen field. You *will* succeed as an electronics student, and you're about to discover how.

The Time to Learn

If you are going to learn electronics, you will need time, all that you can get. Actually, you have as much time as anyone else; after all, a 24-hour day is the same for everyone. What you want, of course, is to get more accomplished in the same amount of time, to use your time more wisely. To do that, you must learn to *manage* your time, to make better use of it. After all, you manage just about everything else, so why not manage your most valued resource—time?

Yet the concept of time management often seems abnormal to many students. Even in a 24/7 world, when the subject is broached, one often hears defensive, even strident, comments:

"It's nerdy to carry around one of those time-management books."

"I don't need to manage my time since I do the same thing every day."

"That time-management stuff is for old people who have trouble remembering anything."

"I don't want to be that organized; I want to be free and loose."

"I tried it once and it doesn't work."

These objections aside, it would seem, given the obligations of a typical electronics technician student today, some type of time management is called for. Take Ralph Lee, for example: "My classes are scattered all over the school between 8 A.M. and noon, 5 days a week. I have English 101, Math 135, Physics 114, Fundamentals of Electronics 112, and Electricity Lab 113. My work schedule is pretty mixed-up, too. Some days I work 8 hours, other days 4 to 6 hours, and a couple of days a week not at all, except when I am covering for a fellow worker. Of course, I have to find time to study. I have a younger brother to take care of, and naturally I want to spend time with my girlfriend. Then there are the little things that come up every day—going to the cleaners, working out, fixing my car. It's hard to keep it all straight. Naturally, I wonder if I am making the most effective use of my time."

If Ralph isn't practicing some form of time management, the answer is probably no, he is not. Chances are, Ralph could use some help.

But what exactly is time management? At its most elemental level, it is simply a method of recording all the things you need to do, of prioritizing the list, and of providing a check-off procedure that tells you when each item is completed. Time management results in at least three clear benefits:

1. By writing down what is to be done, where, and when, you have put it on paper, in black and white. As a result, it "stares you in the face"—it cries out to be "checked off," or accomplished. "Do problems 1–20, on pages 83–85 of the electronics text between 4:00 and 6:00 P.M. in the school library on Thursday, April 4" is a statement that is hard to ignore. It's also quite satisfying when it gets checked off your list as being completed.

2. If you write it down, you can forget it, or take it off your mind, temporarily. This is especially true of so-called little things. Suppose you're in a classroom taking notes as the instructor is discussing Ohm's law. You know the subject is important, and you're paying close attention. All of a sudden you remember that you must purchase a reference book at the student store after class. You resolve to keep that "note" in your head while the lecture proceeds. Your mind is now forced to juggle the professor's information on Ohm's law with your mental note on getting to the student store. You aren't giving Ohm's law the attention it deserves. On the other hand, a quick note written in a *planner book* (Figure 6.1) of some kind would relieve you of the

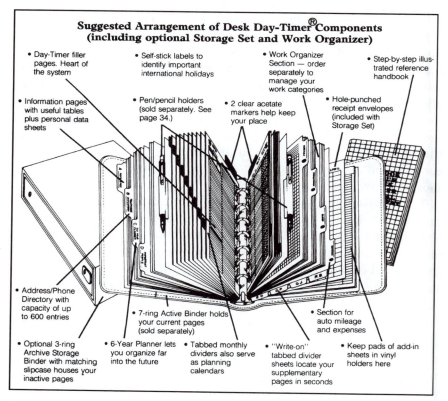

**Suggested Arrangement of Desk Day-Timer®Components
(including optional Storage Set and Work Organizer)**

- Day-Timer filler pages. Heart of the system
- Information pages with useful tables plus personal data sheets
- Address/Phone Directory with capacity of up to 600 entries
- Optional 3-ring Archive Storage Binder with matching slipcase houses your inactive pages
- Self-stick labels to identify important international holidays
- Pen/pencil holders (sold separately. See page 34.)
- 7-ring Active Binder holds your current pages (sold separately)
- 6-Year Planner lets you organize far into the future
- Tabbed monthly dividers also serve as planning calendars
- Work Organizer Section — order separately to manage your work categories
- 2 clear acetate markers help keep your place
- "Write-on" tabbed divider sheets locate your supplementary pages in seconds
- Step-by-step illustrated reference handbook
- Hole-punched receipt envelopes (included with Storage Set)
- Section for auto mileage and expenses
- Keep pads of add-in sheets in vinyl holders here

Figure 6.1
Day-Timers® components.
Source: Day-Timers, Inc.

mental tension that exists and allow full concentration on the electronics subject being presented in class.

3. Planning a week, a month, or even a semester ahead, perhaps using a wall calendar, allows you to see the "big picture," or overall process. Of course, not every detail of your life can or should be planned a semester in advance. The idea is to have an overview of the main activities to be accomplished.

There are numerous books, notebook-type planners, hand-held electronic organizers, seminars, and workshops available to students that adequately address the issue of time management. Check with your school counselor for details. If, like Ralph Lee, you sense that help is needed, investigate the many time-management options. Finding the time to learn is the first step toward succeeding as an electronics student.

Learning Skills for Electronics Students

As an electronics student, you will be taking many courses in electronics, almost all of which focus on the installation, testing, troubleshooting, repair, and construction of electronic devices and systems. In such courses, most learning will occur in the classic classroom-laboratory environment.

In the classroom you will be presented with electronics theory, primarily through your instructors' lectures and the assigned textbooks. You will take notes on what the instructors say, ask questions for subject clarification, enter into dialogue with your instructors and fellow students, and, of course, take examinations. After a class you will read and highlight the text, work chapter problems, and answer chapter questions. In sum, through discussion and reading, you will acquire the theoretical foundation of electronics—a prerequisite to gaining command of the more hands-on activities you will actually engage in as an electronics technician.

In the laboratory you concentrate on proving the concepts learned in the classroom. Usually, you work from a lab manual made up of experiments. In the process of completing such experiments, you learn to use electronic equipment, breadboard electronic circuits, troubleshoot and repair electronic devices, and perhaps even build and install electronic equipment. Furthermore, throughout your lab experience, you learn to interact with future electronics technicians—that is, your fellow students.

To succeed in electronics courses, both in the classroom and the laboratory, you will want to sharpen and hone four critical proficiencies. You must master *reading, listening, mathematics,* and *circuit construction, testing,* and *troubleshooting* skills. Let's take each in turn.

Reading: Sharing Knowledge and Know-How

Before we talk about the "how" of **reading**, let's take a moment to discuss why, if you're a typical technical or vocational student, you're probably not reading as much as you should.

All national surveys show a declining interest in reading. Television and lack of time are given as the main causes. However, beyond such obvious factors, there are more specific explanations as to why vocational students, in particular, seem turned off to reading.

First, there is the "nerd" or "bookworm" factor. Believe it or not, there are still college students who think it's "uncool" to be seen in the company of books. Because reading is a solitary activity, one who engages in too much of it is perceived as antisocial. A news report once described what neighbors thought of an accused rapist. He was spoken of as "a loner, a man who spent a lot of time with his computer, a guy who *read books.*"

Second, for many vocational students, the world is divided into "readers and doers." Such students value experience, and reading is perceived as an attempt to "steal" someone else's expertise, an unworthy, if not impossible, task. Vocational students want to get the job done. How-to books are all right, but reading of a more general nature is often seen as unproductive.

Third, early attempts at reading in the primary grades may have gotten off to a bad start, and it's been a downward spiral ever since. If you read poorly, you won't read. If you don't read, you'll continue to read poorly, and so on. Reading, therefore, becomes a chore, a laborious, time-consuming task, with little comprehension.

These elements, the nerd concern, the readers and doers issue, and the reading difficulty factor, however, must be rejected or overcome if learning is to take place. If you truly desire to know electronics, you must read, read, and read.

Reading is a means to a very important end. You can't continually depend on someone to be there to show you what to do. But bring along a book, and the expert is always right by your side. Shared knowledge and experience is all right, desirable, even required: Where would we be without it? Much of what you do as an electronics technician has already been done by others. Many of these individuals have recorded their experience in the form of books or manuals that anyone can obtain. You'd be crazy not to take advantage of their know-how.

Furthermore, self-serving as it may sound, people that you need to impress will be impressed that you are a reader. This is because your boss or your professor knows that people who read think differently than people who do not. According to College Board President Donald M. Stewart, "The ability to read is linked to the ability to process, analyze, and comprehend information. I guess that's called thinking." Yes, and thinking leads to doing. By being a reader, you will become a thinker as well as a doer.

The *textbook,* of course, is what you will spend most of your time reading. After you have purchased it, set aside 1 hour—no more, no less—to peruse your new source of knowledge. This is a get-acquainted time. Find a comfortable spot and resolve to spend the next hour probing the book that will surely become your single most important source of electronics information during the entire semester.

What should you look for? First, just flip through the pages. Is it a one- or two-color book? Is it filled with illustrations? Are the chapters long or short? Is there a list of objectives at the beginning of each chapter? Are there problem examples? Are there questions and problems at the end of the chapters? Is there an introduction and summary for each chapter? And what about front and back matter—that is, a preface and introduction at the beginning and appendices, glossary, and index at the end? Take the time now to become familiar with your textbook's layout and features.

Next, examine more closely how the book is organized. Are there major parts, or sections? How are the chapters broken down? How are subheadings identified? What you are trying to do is find the author's outline. It's there, hopefully not too well hidden. Discover it, and you'll conquer the text with little effort.

Now zero in on one chapter, any chapter. First, look it over from beginning to end. Get a page count. Determine the number of headings and subheadings, perhaps marking them in the standard outline format: Roman numeral, uppercase letters, arabic numbers, and so on.

Second, read the chapter objectives, introduction, and summary, in that order. Third, read a sampling of chapter questions and problems. Do the same with chapter examples. Fourth, examine closely every figure, in order. Read the captions. Note if a figure is primarily a schematic, a graph, or a mathematical expression. In some textbooks, it is possible to comprehend the entire subject matter, at least at an elementary level, just by reading and studying the figures.

Try to believe that all this perusing is not a waste of time; it's part of the getting-to-know process, an hour well spent.

When you are finally ready to begin reading, it must be earnest, *active* reading. Take off the earphones, turn off the TV, sit at a desk or table (do not lie in bed), and grab a pen and highlighter. Get ready for some serious work.

There are a number of approaches to reading that can be taken at this point. One of the most effective is to be found in a small, inexpensive, 72-page book written by Thomas F. Staton, entitled *How to Study.*

The book's study theme is built around what Dr. Staton refers to as the PQRST method, where P stands for *preview;* Q, for *question;* R, for *read;* S, for *state;* and T, for *test.* The idea is to *preview* the assignment before reading it, *question* what it may include, *read* it, using your eyes and your brain, *state* in your own words what you have read, and *test* your memory of it a few hours or days later.

The PQRST steps must be done while you're in the active, not passive, mode. You cannot lounge back, with textbook in hand, and just let your eyes scan the printed page. You must sit up, concentrate on the matter at hand, and engage your brain in the active process of learning. You have to probe and dig into the reading material, jotting margin notes whenever appropriate. Yes, such active reading is more difficult than lying back and listening to the stereo while flipping pages. But it is also much more effective. And that's the bottom line. It's not how much time you spend reading, but rather what you understand and retain as a result of the studying process.

Listening: Getting More Than Just the Word

While learning demands reading, it also necessitates listening. In fact, between lectures and multimedia presentations, you're likely to spend three times as much time listening as you do reading. That may be just fine with some of you; however, there is a problem: Good, effective listening is the hardest of all study skills to master. This is because unlike reading and seeing, listening involves another person, the speaker. As a result, you must coordinate your thinking with that of the speaker, which is not always an easy task. But effective listening is a skill that can be learned, and to discover how, let's closely examine the teacher as a lecturer and your role as an active, engaged listener.

As a lecturer, your instructor must determine what is most important for you to know and clarify, through speech, discussion, and demonstration, concepts that you may have trouble grasping. For learning to take place, however, you need to enter into a dialogue with your teacher. Through active listening and questioning, you and your fellow electronics students can turn what might otherwise be a traditional and boring lecture into a discussion of interesting facts and concepts. Most teachers want this to happen, but they need help from students, in the form of feedback, to make it work. (One survey found that on the average, the classroom teacher asks 90 percent of the questions and answers 80 percent of them. In other words, professors are talking to themselves.)

But how do you make use of your teacher in the classroom? You do this mainly by asking questions. Don't be afraid of interrupting the teacher or wasting the class's time. And don't be concerned if yours is a so-called dumb question. If asking the question clarifies an issue for you, raising your hand is the smartest thing you can do—not only for yourself but also for your fellow students, who were probably just as perplexed as you.

Although asking questions and providing feedback to your instructor is critical to enhancing your learning in the classroom, the truth is, most of the time your instructor will do the talking and you'll do the listening. If that is to be the case, how can you make your listening more effective? Here are some suggestions:

- If only your ears are at work when listening, knowledge will not be gained; the mind must also be at work. When your instructor or a fellow student speaks to you, concentrate, think, and reason with regard to what is being said.
- Look for the general plan or theme of the lecture. Think about how it relates to the whole, what you heard yesterday, and what you are likely to discover tomorrow.
- Go beyond mere facts to concepts. Yes, in an ac circuit, capacitive reactance is inversely proportional to frequency, a fact that can be stated in a simple formula. But why is this so? What is the underlying principle? That principle, more than the formula, is what you should be after.
- When your professors lecture, listen for special emphasis. In a number of ways they will indicate what is most important. For example, do they write a name, formula, phrase, or concept on the chalkboard? If so, you should note it too. Do they say, "This is an important point?" If so, deal with it as such. And, the phrase *note the following* should be treated literally.

Speaking of note taking, here are a few hints that will make yours more successful:

- Record only the high, or key, points. You're not a stenographer; you don't need to take down every word.

- Record what is being said in your own words. Rephrasing the lecturer's words will help you learn and remember the maximum amount from the lecture.

- Go over your notes after class. Use the time to clean them up a bit and to insert additional ideas and clarifying words.

Effective listening, so important to the learning process, requires lots of practice. Look around; you'll find plenty of opportunity to practice. Practice on your fellow students, members of your family, friends, and, of course, your professors. A greater understanding of any subject you study will be the result.

Mathematics: Electronics Theory the Easy Way

If you're like many students, you knew the subject of mathematics would come up sooner or later, and you hoped it would be later, or not at all. The fear of math probably deters more students from going into electronics than anything else. For many, their reasons are as follows: "There's a lot of math in electronics; math is difficult; I wasn't particularly good at math in high school; therefore, I can't succeed in electronics." This logic, or lack of it, aside, one point is worth noting. From the standpoint of an electronics technician, the math involved, at least in the core courses, is relatively straightforward and not nearly as complex as you might have been led to believe. Sure, you should take all the advanced math you can: algebra, geometry, trigonometry, even calculus. You can never be too numerate. But for the most part, your math, at the first-year level, will consist of manipulating formulas and interpreting graphs. We will, therefore, in the limited space available, concentrate on demystifying these two subjects. But before we do that, let's take a moment to explore a tool that will make working with formulas and graphs a lot easier, the indispensable **electronic calculator.**

A hand-held electronic calculator is probably the most essential piece of equipment that you, as an electronics student or working electronics technician, will ever own (Figure 6.2). Today, for less than $20 you can place in your palm the power of a mainframe computer of 25 years ago. Such a calculator contains all the mathematical functions you are ever likely to use. The calculator and its operating manual should go everywhere you go, to the classroom, laboratory, and place of study. It's difficult to imagine learning electronics today without it.

Figure 6.2
An inexpensive scientific calculator.
Source: Radio Shack.

Electronic calculators have changed the study of electronics at all levels in at least four significant ways. First, with a calculator, you can deal with real-world problems rather than those that have been rounded off to make calculations easier. Finding the parallel resistance of a group of 3.3, 4.7, 6.8, 10, and 22 kilohm resistors is no more difficult with a calculator than doing the same thing with 2, 4, 8, 16, and 32 kilohm resistors. While finding the parallel resistance of the latter, non-standard group of resistors would be easy enough using longhand, you wouldn't even want to try the first, "real" group of resistors without your trusted calculator.

Second, there are functions available on an ordinary $15 calculator that would take days or weeks to work out by hand. The logarithmic and trigonometric functions come quickly to mind.

Third, with a calculator, press the right buttons in the right sequence, and you're going to get the right answer every time. Furthermore, that answer will be accurate and carried out far to the right of the decimal point.

Fourth, using a calculator will reduce or, in many cases, eliminate many of the little mistakes associated with laborious, multistep calculations.

For these and many other reasons, from the moment of purchase seek to make your calculator a trusted friend and invaluable companion. Every day, strive to do something more with it, to conquer new functions and procedures. Doing so will make your involvement with mathematics much more successful and a whole lot easier.

Now, for the *formula*. A **formula** is simply a mathematical equation that shows you how to find something (an unknown) if you know the ingredients and their relationship. Coca-Cola keeps the formula for Classic Coke a guarded secret. But if you knew the ingredients (malts, flavorings, spices, etc.) that they used and their relationship to each other, you'd have something quite valuable. It's the same with electronics. Consider the formula

$$X_C = \frac{1}{2\pi f C}$$

Essentially, we ask, What is the formula "saying"? That is, we want to know what the ingredients are and what their relationship is.

The preceding formula says if we want to find capacitive reactance, X_C, in a circuit, that circuit must produce (contain) a sine wave, 2π; ac voltage, f; and some capacitance, C. The letter f tells us the circuit will have to have an ac source, the constant 2π (approximately 6.2832) says that the source must be sine-wave ac, and C tells us capacitance, in farads, has to be present. Those are the ingredients of the circuit.

Since the formula also says X_C is equal to 1 divided by the quantity 2π multiplied by f and C, we have a reciprocal, or inverse, relationship. Thus if frequency (f) goes up, X_C goes down and vice versa. The same is true for capacitance, C.

Now, to solve the equation, all we have to do is fill in the values for f and C, perform the indicated arithmetic operations in the correct sequence, and presto—we have the value of X_C.

The beauty of the formula, indeed of all mathematics, is the straightforward way in which it communicates relationships. Look at Ohm's law expressed as a mathematical equation:

$$I = \frac{V}{R}$$

In words, this equation means current is proportional to the applied voltage and inversely proportional to the resistance. Or, for a fixed resistance, the greater the voltage (or pressure) across a resistor, the more the current, and the more the resistance for the same voltage, the less the current. After reading that, don't you long for the simplicity of the formula?

Try not to dread mathematics; look forward to it as the *easy* way to solve problems. Learn what a formula says and what it means. Do so and you'll overcome any math phobia that you may still have.

Finally, let's turn to a discussion of *graphs*. **Graph** is short for graphic (pictorial) formula. We can say it's a way of representing mathematical information in pictorial form. As such, you would rightfully expect the graph to find considerable use in electronics.

Graphs tell a story. Once you have discovered, read, and analyzed the story, you have interpreted the graph. Let's take an example.

Figure 6.3 shows a typical graph of the kind you're likely to find in an electronics text. Upon first seeing it, you might simply lean back and look it over. In the same way as you perused the textbook, peruse the graph. Don't try to analyze it just yet; just examine what it contains.

Next, look closely at the two axes: the horizontal and the vertical. Graphs usually plot one variable against another; in this case, frequency, f, against reactance, X. Note that frequency, represented on the horizontal axis, increases to the right as we get farther from the vertical axis, and that reactance, on the vertical axis, increases as we move upward from the horizontal axis. Once we discover what is being plotted (frequency against reactance), we usually have half the story.

Now, we fill in the details. First, the "story" tells us that inductive reactance, X_L, increases with an increase in frequency in a linear manner

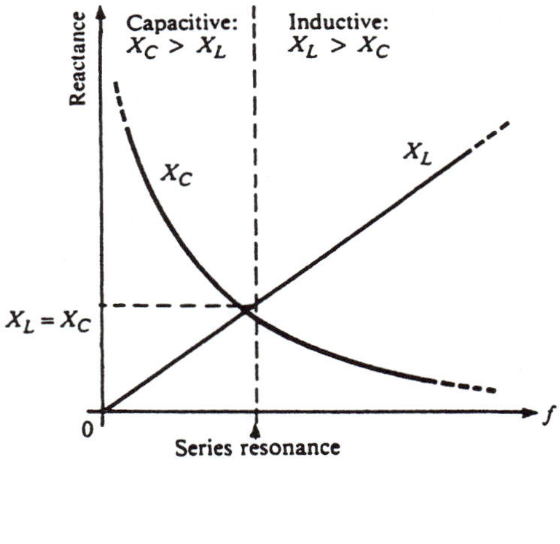

Figure 6.3
A typical graph.
Source: Thomas L. Floyd,
Electronic Fundamentals: Circuits,
Devices, and Applications, 2d ed.
(Upper Saddle River, NJ: Prentice
Hall, 1991), p. 554.

(straight line). On the other hand, capacitive reactance, X_C, decreases with an increase in frequency and not in linear fashion (note the curve).

Continuing on, we note that X_C and X_L cross at one point. At that point the two are equal, or $X_C = X_L$. From their point of intersection, a vertical dotted line is drawn to intersect the horizontal axis. This intersection represents what is known as the series resonance frequency.

Finally, we note that with frequencies to the left of the dotted vertical line (lower frequencies), X_C is greater than X_L ($X_C > X_L$). To the right of the same line, X_L is greater than X_C ($X_L > X_C$).

There you have it. Without even knowing exactly what reactance and frequency are, you know what the graph is all about.

Circuit Construction, Testing, and Troubleshooting: "Doing" Electronics

It's time now to turn our attention to the hands-on skills needed to succeed both as an electronics student and a working technician. In the laboratory portion of your courses and later on the job, you will spend a great deal of time constructing, testing, and troubleshooting electronic circuits. Following are a few observations with regard to each activity.

In **constructing** electronic circuits (and testing and troubleshooting them, for that matter), you must, above all else, know how to read a schematic drawing (Figure 6.4). The schematic is a construction guide, as are architectural plans for builders and orthographic projections for machinists. The schematic consists of two elements: (1) symbols representing electronic components, and (2) lines that stand for circuit traces (or wires) that connect the components. Together, these symbols and lines represent an electronic circuit.

When first encountered by the neophyte electronics student, all these strange symbols with connecting lines in a schematic can be quite intimidating. Yet during my many years as a junior-high, senior-high, and community-college instructor, I have seen students learn to read schematics with remarkable ease in a relatively short period of time. This is because, I suspect, they soon come to realize what experienced techs already know—complex schematics differ from simple ones only in that the former have more symbols and lines, not because they are qualitatively more involved. If you can read 3 sentences in this book, you can read 15. It is the same thing with schematics. If you can understand a drawing with 3 transistors, all that's required to read one with 15 transistors is a little

Figure 6.4
A typical schematic.
Source: From *Digital Electronics through Project Analysis* (p. 15), by R. A. Reis, 1991. Upper Saddle River, NJ: Prentice Hall. Copyright 1991 by Prentice Hall Publishing Company. Reprinted by permission.

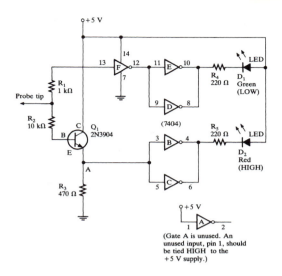

more time. In other words, once you have mastered the "language" of component symbols and connecting lines, you're ready to tackle any schematic. You will want to improve your "vocabulary" of schematic symbols, just as you must increase your knowledge of new words in your daily reading. But by starting with simple schematics and progressing to more complex ones, noting carefully each new symbol as it is encountered, you'll be reading schematics like a pro in no time.

In the school lab (and, later, on the test bench or out in the field) you are going to spend a lot of time **testing** circuits. To do so successfully will require, first and foremost, gaining knowledge and experience in the use of test equipment. In some cases the effort will be minimal, a minute or two to become competent in the use of a continuity checker. On the other hand, it can take months of intensive study and application to master the operation of a microwave network analyzer. But regardless of equipment complexity, to become truly proficient in its use, you must conquer the intimidation all such equipment can foster, at least when it is first encountered.

For example, when confronted with a moderately complex oscilloscope, all those dials, knobs, switches, and waveforms can scare the heck out of you. It is intimidating, no question about it. Yet, as an experienced tech has suggested, "To ease your anxiety, try thinking of the vast array of controls in the same sense as the controls on a stereo's graphic equalizer. Each has a specific function, but once it's understood, it's no big deal. Just set the controls to your liking (so you can see the particular waveform the way you'd like). You don't need to tweak controls continuously. Soon you'll be amazed at how fast you get the hang of using the scope."

In other words, you probably already know how to operate some fairly sophisticated electronics equipment—a graphic equalizer, a home computer, maybe even your VCR. If you can learn to operate such equipment, you can master the operation of electronic test equipment too. Just get out the operator's manual, take your time, and go to it.

Finally, let's briefly turn to **troubleshooting,** possibly the single most important activity that you, as an electronics technician, will engage in. More than anything else, your success as a troubleshooter depends on your having the correct attitude. You must realize that wherever you go, whatever you do, trouble and failure will surround you. It is you, the electronics technician, who is called in when something doesn't work. It is you who must determine the problem and fix it—often in short order and under considerable pressure. If you are uncomfortable with this sort of thing, if you get unduly frustrated when things don't work right, if you are extremely impatient, you might want to think twice before becoming an electronics technician. Stop for a moment and take this simple self-test. Assume that you have just completed an electronic kit project in the lab. A fellow student in the class has done the same. You both plug in your projects, and guess what? The other project works; yours doesn't. Who's better off, you or your classmate? In a real sense, you are. Now in addition to learning how to build a project, you'll learn how to troubleshoot and repair it as well. Without being too facetious, if you can honestly feel that way, you're going to make a great electronics technician.

Grade-A Learning

In this final section, we explore what you can do to heighten learning and get the best grades possible in your electronics classes. We look at study strategies, use of the learning resource center, the help computer-assisted instruction can provide, and how to prepare for and take the kind of examinations you're likely to encounter in electronics courses. We conclude with some thoughts on the grading process.

Although you will spend considerable time in the classroom and laboratory in the pursuit of your electronics education, you are sure to spend even more time in outside study. One of the best **study strategies** to adopt is one that involves plenty of *group study,* where four or five students meet regularly in a quiet place to exchange ideas and think out, talk out, work out, and write out fundamental concepts in electronics. In such a setting it is important that everyone be on time, be prepared, and contribute to the discussion. By doing so all participants benefit in at least three ways:

1. In a study group of fellow electronics students, you have three or four tutors right at your table. You can get quick, thoughtful answers to questions that require further explanation.
2. When you're in the tutor's role yourself, explaining and clarifying concepts for others, you also learn a great deal. Any teacher will confirm that teaching others in a clear manner enhances your own learning.
3. By participating in a study group, you are learning to work in groups, something you're likely to do a great deal of later while on the job.

An excellent place to do your studying, as an individual or group, is in what used to be called the library but is probably known as the **learning resource center** (LRC). The name change is significant. Calling it the LRC means that more than books are at your disposal; it means that the full range of learning materials are there to aid you in acquiring knowledge. Of particular interest to electronics students are the audio and visual aids (audio and videotapes), and, of course, the Internet. The same tapes that your instructor shows in class are most likely available for your individual viewing in the LRC. Take full advantage of these tapes and all that the LRC has to offer. Make the learning resource center the centerpiece of your learning strategy. You have paid your tuition; you have a right to all the services the college and the LRC provide.

Another study aid you're likely to find in the LRC or your electronics department is a computer with multimedia **computer-assisted instruction** (CAI). Such instruction comes in the form of software programs that can aid in circuit design and understanding. CAI offers a number of advantages for anyone wanting to learn new concepts in electronics:

1. Animated graphics is usually an integrated feature.
2. Dynamic current flow is easy to depict.
3. Spreadsheet-type changes, where a change in one variable affects all others, are commonly incorporated.
4. Automatic scoring and instant feedback are a part of almost all CAI programs.
5. With CAI, you proceed at your own pace.

If those aren't adequate reasons, keep in mind that getting involved in CAI gets you involved in computers. Never pass up the opportunity to do that.

All this studying and learning aside, for most students the moment of truth comes during **test-taking.** In electronics, as opposed to history, for instance, you're likely to take many tests of the objective type rather than one or two (midterm and final) essay exams. That's probably good news to some of you. However, if you're like many students, while you know you are just as smart as the next person, you have considerable trouble preparing for and taking examinations of any kind. If preparation is a concern, here are a few tips:

- Find out as much about the test as you can from your instructor. What type of test is it going to be? How many items will be on the test? Will the test cover material from the lectures, textbook, and lab work, or all three?
- Make up your own questions and problems ahead of time.
- Get to bed early the night before an exam.

When taking the exam, keep these points in mind:

- On a multiple-choice exam, read through *all* the choices first.
- On true-false questions, if you can think of *any* exception to a positive statement, the answer is false.
- Don't look for trick questions; instructors are not out to get you.
- First, answer the questions you are sure of. Doing so will give you confidence, and the association may help you with the more difficult questions.

Finally, try to avoid panic. Arrive for the test early, don't swap information at the last minute, peruse the entire test when you first receive it, and listen carefully for any last-minute instructions and comments your instructor may have.

In the end, of course, there is the issue of **grades.** Grades are important. It is expected that you will earn an A or a B in courses within your major. Even so, more and more 4-year colleges and employers are looking at what are known as *multiple measures* when evaluating a student's transferability or employment prospects. Multiple measures say: "Consider how the student's lifestyle contributes to his or her grade average in evaluating the applicant's prospects for success." What other obligations, beyond school, was the student burdened with while attending classes? For example, if you're a single mother of three, working 40 hours a week, taking a full load at a local community college, and maintaining a 2.5 grade point average, all those other factors should be considered when determining eligibility for a 4-year college or a job as an electronics technician. In other words, such factors as earning requirements, family commitments, and language problems do play a role in your overall success at school.

Nonetheless, do not interpret the preceding comments to mean that your knowledge of electronics, reflected in course grades, is unimportant. When seeking employment as an electronics technician, you will be expected to have acquired a knowledge of electronics. Grades and faculty recommendations say a great deal about that knowledge or the lack of it.

Summary

In Chapter 6, we began by taking time to learn about time—how to use it more effectively. Next, we looked at learning skills for electronics students. In particular, we saw how to master reading, listening, mathematics, and circuit construction, testing, and troubleshooting skills. From there, it was on to specific study approaches, with emphasis on how to use the learning resource center and multimedia computer-assisted instruction. We concluded with a look at various test-taking approaches.

Review Questions

1. At its most elemental level, _____ _____ is simply a method of recording all the things you need to do, prioritizing the list, and providing a check-off procedure that tells you when each item is completed.

2. The ability to _____ is linked to the ability to process, analyze, and comprehend information.

3. Through _____ listening and questioning, you can turn what might otherwise be a traditional lecture into a discussion of interesting facts and concepts.

4. A _____ is simply a mathematical equation that shows you how to find something if you know the ingredients and their relationship.

5. _____ is short for graphic (pictorial) formula.

6. The _____ drawing consists of two elements: (1) symbols representing electronic components and (2) lines that stand for circuit traces (or wires) that connect the components.

7. To test a circuit, you must, first and foremost, gain knowledge and experience in the use of _____.

8. _____ is probably the single most important activity that you, as an electronics technician, will do.

9. One of the best study strategies to adopt is one that involves plenty of _____ study.

10. On true-false questions, if you think of _____ exception to a positive statement, the answer is false.

Individual or Group Activities

1. Obtain a time management-type planner book and use it for a month to organize your daily activities. After the 1-month period, evaluate its success in terms of your getting things done *and* the peace of mind it may have engendered.

2. Form a study group of four or five students. Set some ground rules with regard to individual preparation and participation. Plan to meet regularly and frequently. After a 1-month period, evaluate the group's success in terms of knowledge gained, improvement in test scores, working with others, and gaining self-confidence.

3. Resolve to practice your listening skills. Apply the techniques discussed in the chapter when listening to classmates, teachers, friends, and fellow employees. After a 1-month period, give yourself an honest assessment. Have your listening skills improved? How do you know?

4. Learn to read schematic drawings. Start with simple schematics and projects. Use a colored pencil to trace over a schematic drawing as you compare it with an actual circuit. Work up to more complex projects and schematics. As you gain expertise, give yourself a self-test with a really complex drawing.

Issues for Class Discussion

1. Discuss and share test-taking strategies with your classmates and instructor. Have your teacher explain how he or she prepares an examination, what he or she looks for in answers to the questions, and so on.

2. Discuss student reading habits. Do you think students who read think differently from those who do not? Is the nerd factor for real, or is it just that you don't have the time to read more? What can be done to improve student reading habits?

3. Discuss your math phobias with the rest of the class. Is fear of math real? If so, what can be done to reduce this fear? Are math tutors available? Is anyone in the class prepared to act as a math tutor? If so, can a list of such tutors be distributed?

4. Have your instructor discuss his or her grading procedure. What factors determine a given letter grade? Does the instructor take into consideration such things as class attendance, tardiness, class participation, and helping fellow students when determining a final grade?

5. Evaluate your textbook with fellow students. What are its strengths and weaknesses? How does one take advantage of its strengths?

6. Evaluate the PQRST method of study. Share your results with the class.

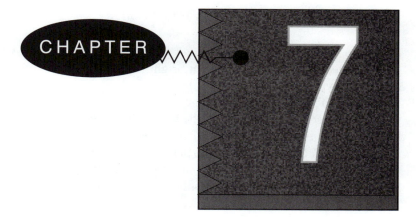

Being an "Electronics Activist":
Electronics Beyond the Classroom

Objectives

In this chapter you will learn:

- How building electronic projects can be a confidence booster.
- That electronic kits provide a convenient way to do hands-on electronics.
- The advantages of joining an electronics club.
- How ham radio is more than just a hobby, it is a career opener.
- How you can become certified as an electronics technician.
- The advantages of attending trade shows and conventions.
- How to make your summer vacation a time of opportunity.
- How an information interview can lead to a job.

It is 5 years from now. You are working for ABC Electronics, Inc., supervising six electronics technicians. You have been given the go-ahead to hire a new tech, and you have narrowed the choice down to two candidates, Larry and Bob. (In this book, Larry and Bob are fictionalized characters.) Both are recent graduates of 2-year colleges, with A.S. degrees in electronics, and both seem bright and eager for the job. Which one should you hire?

Larry has a 3.2 grade point average (GPA), his résumé is in order, and he presents a nice appearance during the interview. He knows the electronics he has learned in school and is well grounded in electronics theory. However, when the conversation turns to the practical aspects of electronics, the actual building, testing, installing, and troubleshooting of electronic devices, Larry seems ill at ease; his confidence begins to wane. As he admits, "I don't have much experience in electronics beyond the classroom."

Bob also has a good GPA, 3.1. And, like Larry, he knows electronics theory. But when you begin to discuss the actual doing of electronics with Bob, instead of sinking into his chair as Larry seemed to, Bob straightens up and brightens up even more. From his briefcase emerges a completed frequency counter project that he built from scratch, on his own, after school, from plans he found in *Poptronics*. He designed and fabricated the PC board, installed all the components (the soldering is first rate), built the chassis and cabinet, and did all the necessary testing and troubleshooting on his project. Bob is clearly excited and confident; he can't stop talking about his project and what he learned from building it. His enthusiasm is contagious, and you can't help but be impressed.

Bob continues on about the electronics club he joined, the field trips he has taken, the certification he has obtained, and the electronics trade shows and conventions he has attended. Bob has clearly taken his electronics studies beyond the classroom, and it shows in his self-confident attitude.

Which one would you hire? The answer is obvious. By becoming an "electronics activist," Bob has clearly set himself apart. He has the edge— and he gets the job.

In this chapter we show you how to become a little more like Bob— how to gain confidence by reinforcing your study of electronics after class. First, we look at project building as a confidence builder. Next, we examine how organizations and clubs can extend your awareness and knowledge of electronics. Then we move on to the "ultimate" electronics hobby, amateur radio. We discuss certification for electronics students and technicians. We talk about the advantages of attending trade shows and conventions. We examine ways to make your summer vacation truly productive. And, finally, we explore the "information interview" as a means of learning about industry and preparing yourself for getting that first job as a full-time electronics technician. As you'll see, it's never too late to become an electronics activist.

Project Building: A Confidence Booster

Over my many years as an electronics instructor, I have seen nothing that gives a student more knowledge, pride, confidence, and a greater sense of achievement in electronics than project building. Why is this so? First, students know that the skills of schematic reading, PC-board design and fabrication, soldering, cabinet making, and testing and troubleshooting acquired in project building are the very skills that will be demanded of them as electronics technicians. Second, students also come to understand that the project is the one place where the subdisciplines of electronics can be joined—where digital, analog, communications, industrial control, and other fields of electronics can find a common system to populate. Thus, for both reasons, building, testing, and troubleshooting your own electronic project gives you the best possible exposure to the wide world of hands-on electronics.

For the typical electronics student, there are three approaches to project construction: (1) purchasing an electronic kit, (2) building a project from directions supplied in an electronics magazine, or (3) selecting a project from the hundreds of how-to electronics books available. We'll take a look at all three approaches while also discussing the creation of a working environment and the purchase of necessary tools and equipment.

Electronic kits, where materials and assembly instructions are supplied, range in complexity from a simple continuity checker to a microprocessor-controlled robot. A number of companies are ready to supply such kits, ranging in price from less than $10 to more than $300 (though the average is about $20). See Appendix 1 for a list of companies making these kits. Write to the various companies and ask for their latest catalog. Just tell them you're an electronics student who wants to get into kit building.

Some of the projects these companies present are fun, even whimsical. Chaney Electronics, Inc., offers such items as a mosquito repeller kit, rolling dice kit, heartthrob kit, and an electronic minefield game. Kits such as these are great for beginners and make excellent gifts at holiday time.

In a more practical vein, HobbyTron supplies such kits as a digital slot machine, combination lock/alarm, 2-watt audio amp/intercom, and a 0–15 volt power supply, just to name a few. See Figure 7.1.

With Marcraft Corp., you can choose from analog and digital multimeters, AM and FM radios, and a radio-controlled model race car.

Building projects from kits offers a number of advantages. First, you're assured of getting a project whose design has been verified. Second, you will receive the correct parts and components. Third, most kits come with

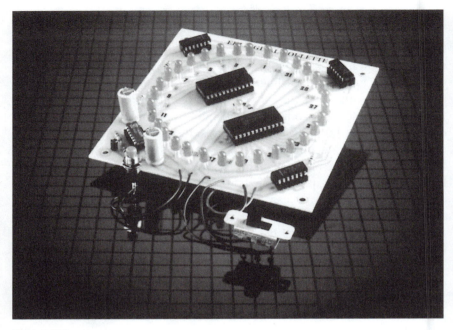

Figure 7.1
Digital roulette wheel project.
Source: Ronald A. Reis, *Electronic Project Design and Fabrication,* 2d ed., Merrill, an imprint of Macmillan Publishing Company, 1992, p. 309.

complete, easy-to-follow assembly instructions. And fourth, if you can't get your project to work, many kit companies will provide troubleshooting assistance over the phone, or you can send the project to them for repair, usually for a nominal charge. If you are just starting out in project building, kits are a good way to go.

If you are ready for a more independent challenge, however, you might examine the latest issue of an **electronics magazine,** some of which are listed in Appendix 2. All articles tell you what the project does, how to construct, test, and troubleshoot it, and, in many cases, where to purchase the necessary parts and electronic components.

Project books represent another approach. As a rule, they tend to supply more detailed construction information than magazine articles. Such books are more narrowly focused in their subject matter. Just glancing at my bookshelf, I notice *Electronic Music Projects, Building and Installing Electronic Intrusion Alarms,* and *Simple Low-Cost Electronics Projects.* Project books can typically be found at electronics stores.

So, you're now convinced that project building is a great idea. But what about the tools, the equipment, and a place to work? That's a fair question. Let's consider the items in reverse order.

It would be great if you had a den, garage, or basement in which to set up your own electronics lab. But it's not necessary. You can do all your work right on the kitchen table.

Next, let's consider equipment. For starters, an inexpensive (around $50) digital multimeter is all you need. Later you can move on to oscilloscopes, frequency counters, and the like.

As for tools, less than $35 will get you a soldering iron, needle-nose pliers, dykes, a screwdriver set, and soldering plunger. That's all you'll need in the beginning.

We need to mention one more thing before you get started. Remember, project building should be, above all, a learning experience. Don't be surprised or disappointed if all does not go exactly as planned or if your project doesn't work the first time you apply power. As an electronics technician, project or device failure will be the story of your life. Now is the time to learn how to handle such headaches.

Why Go It Alone: Electronics Organizations and Clubs

Project building and other after-school electronics activities do not have to be done in "solitary confinement"—far from it! You can be an electronics activist with a colleague or a whole group of colleagues. When it is the latter, your involvement is usually through a campus club or organization. Being a member of such a group offers many benefits. Here are just a few:

- Provides you with an opportunity to make friends.
- Allows you to network for job opportunities.
- Lets you exchange ideas with other electronics enthusiasts.
- Lets you receive (and give) help with school work.
- Gives you a chance to give and receive moral support.
- Provides you with fun and recreational opportunities.
- Looks good on a résumé and shows you can get along with others.
- Gives you a chance to develop leadership skills.

- Gives you a chance to develop speaking skills.
- Provides you with an opportunity to hear guest speakers.
- Lets you develop contacts with industry.
- Allows you to visit technical facilities.
- May offer summer job opportunities.
- Provides an opportunity to develop competitive student projects.
- Gives you a vehicle to help promote your electronics department on and off campus.
- May offer opportunities to purchase magazines, books, and electronics parts and equipment at a discount.

Generally, such campus clubs fall into one of two categories. Some student clubs or associations are part of regional or national electronics or engineering associations. Most notable are the student chapters (or affiliates) of the Institute of Electrical and Electronics Engineers (IEEE). Other clubs are local, or ad hoc, aligned with no other organization. At Los Angeles Trade Technical College, students have a club called the Electronics Associates Club.

No matter which type of organization your department has or is yet to start, remember this: The club will only be as strong as its members are active. Participate, do your share, get involved—or the benefits listed earlier will never be realized.

Amateur Radio: More than Just a Hobby—a Career Opener

Amateur radio operators, also known as hams, have been "doing their thing," two-way licensed radio communications for enjoyment, not profit, since the turn of the century, when radio was in its infancy. Today, there are hundreds of thousands of "ultimate" electronics activists, ages 6 to 96, from all walks of life and from all over the world. They communicate via voice, digital computer, video images, and even Morse code. They talk through amateur radio satellites using the most sophisticated electronics communications equipment. The majority of hams have such equipment set up in their "shacks" at home, at school, and even at work. For most, amateur radio is not only an exciting hobby that puts participants in touch with fellow hams everywhere, but it is also a great vehicle for expanding their knowledge and skills in communications electronics.

That last point is a crucial one. If you are contemplating a career in some aspect of electronics communications, the electronics knowledge you will gain by studying for the various amateur licenses, working with radio equipment, and talking with fellow hams is invaluable. In addition, having a ham license lets others know, in an official, recognized way, that you possess a good understanding of radio transmission and reception. Furthermore, keep in mind that there is a whole group of licensed hams out there working in the communications industry. When a fellow ham applies for a job as an electronics technician, members of that group can't help but look favorably on such a candidate.

To get started in amateur radio, you will need a license and a station. Today, you can choose from two entry-level licenses: the novice class license and the no-code technician class license. The novice license requires that you pass a 30-question exam covering very basic electrical principles, on-the-air operating procedures and rules, and a *five-word-per-minute Morse code exam.* With the technician class license, you must pass a 55-question exam covering the same principles, but there is *no code exam.*

As for the radio station, most hams have a set-up, or shack, right in their home. However, even hand-held radios are now available, so you can take your station anywhere that you go.

If all this "ham talk" sounds interesting and worth investigating, get in touch with the American Radio Relay League (ARRL) at 225 Main Street, Newington, CT 06111 (860/594-0200). The ARRL promotes the advancement of the amateur radio service. It can supply you with a wealth of information on how to get started in a great hobby—one that just might advance your career prospects as well.

Technician Certification: Much Valued Recognition

Today, electronics technicians are not, for the most part, licensed. Up until recently, the Federal Communications Commission's (FCC) *First Class General Radio-Telephone License* was a requirement for those working as technicians in the broadcast industry. While an FCC license is still offered and desirable to have, in most situations it is no longer required. (However, some companies, such as the SCRTD, use the FCC requirement as an "entrance exam" that they don't have to administer.)

Although licensing in electronics may be fading, **electronics certification** is a growing phenomenon. (See Chapter 8 for a complete discussion of

IT certification.) Its purpose is to distinguish the highly skilled and knowledgeable technicians from those with less experience. Many organizations encourage—and some require—their technical employees to be certified.

There are a number of private-sector certification organizations, two of which are prominent. Both the Electronic Technician Association (ETA) and the International Society of Certified Electronics Technicians (ISCET) offer certified electronics technician (C.E.T. and CET, respectively) certificates. The ISCET, through its exam program, has certified over 46,000 technicians. Once you become certified by either group, you are eligible to join the parent organization and receive all its benefits: discounts on magazines, books, and technical materials and the opportunity to attend conventions and technical seminars (Figure 7.2).

What does all this mean to you, an electronics technician student? Just this: The ISCET, for one, has two levels of certification, **associate level** and **journeyman level.** The first is designed for electronics technicians with less than 4 years experience and (note) *students studying electronics* (Figure 7.3). The 75-question, multiple-choice exam covers the following subjects:

Figure 7.2
Technicians must always update their skills.
Source: International Society of Certified Electronics Technicians.

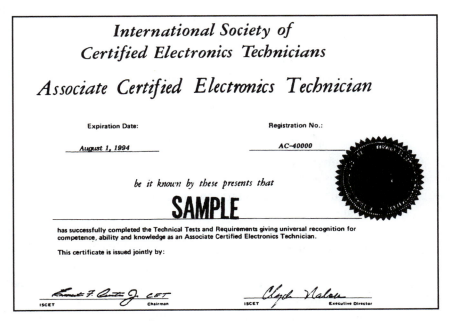

International Society of
Certified Electronics Technicians

Associate Certified Electronics Technician

Expiration Date: Registration No.:

August 1, 1994   **AC-40000**

be it known by these presents that

SAMPLE

has successfully completed the Technical Tests and Requirements giving universal recognition for competence, ability and knowledge as an Associate Certified Electronics Technician.

This certificate is issued jointly by:

ISCET   Chairman   ISCET   Executive Director

Figure 7.3
ISCET associate-level certificate.
Source: International Society of Certified Electronics Technicians.

Basic mathematics

DC circuits

AC circuits

Transistors and semiconductors

Electronic components

Instruments

Test and measurements

Troubleshooting and network analysis

The exam must be passed with a score of 75 percent or better.

The journeyman-level certification is for experienced technicians. You take the associate CET exam plus one or more of several available journeyman options. Here is a list of the options with a description of each, from an ISCET brochure:

Audio: This test covers turntables, tape, disc, and radio. It consists of sections on both digital and analog circuitry, amplifiers and sound quality, system setup, speaker installation, and troubleshooting of audio systems.

Biomedical: Electrical safety, accuracy of calibration for electromedical instruments, and the need for prompt, in-house electronic service are the priorities of this option. The applicant is expected to demonstrate a familiarity with the basic concepts and vocabulary of instrumentation, telemetry, measurements, differential amplifier, and operational amplifier applications.

Communications: Two-way transceivers and the servicing of these types of equipment is examined. Included subjects are receivers, basic communication theory, transmitters, deviation, sensitivity, quieting, and troubleshooting.

Computer: This test covers the operation of computer systems, with greatest emphasis on the hardware. Areas include basic arithmetic and logic operations as related to computer theory, computer organization, input and output equipment, and memory/storage. The technician should have some knowledge of software and programming, and be able to explain troubleshooting procedures.

Consumer Electronics: Subjects covered in this option include antennas and transmission lines, digital and linear circuits in consumer electronics, servicing problems on televisions and VCRs, use of test equipment, and troubleshooting consumer products.

Industrial: The industrial exam covers such items as transducers, switches, power factor, differential amplifiers, basic logic circuits and functions, and elements of numeric control. Some of the equipment covered includes items and circuits more common to the industrial field, such as closed loop feedback, thyratrons, and SCR control.

Radar: General knowledge of both pulse radar and continuous wave radar operation is necessary for taking this journeyman option. This test covers transmitters and receivers, CRT display systems and their power supplies, antennas, transmission lines, and their characteristics.

Video: The rapidly growing field of video is covered by this exam. The technician needs to know NTSC standards, video basics, test signals, and the operations of both the electronics and mechanical systems in video cassette recorders. Also covered are 8mm video, camcorders, cameras and monitors, and the microprocessors used in video operations.

A word of caution with regard to these exams. Don't underestimate their difficulty. Of the 6000 who apply to take ISCET exams each year, only 30 percent pass—it's not an easy test.

With sound preparation and study, however, you can be one of those who do pass. Think of what having such certification now, as a student, would mean on your résumé. Think what joining a parent organization can do in terms of gaining contacts in the industry. It's all worth considering. For more information, contact the ISCET at 3608 Pershing Ave., Fort Worth, TX 76107-4527 (817/921-9101), the ETA, 602 N. Jackson Street, Greencastle, IN 46135 (800/359-6706), or the Computing Technology Industry Association, 450 East 22nd St., Suite 230, Lombard, IL 60148-6158 (708/628-1818, Ext. 301).

Attending Trade Shows and Conventions: Getting Out and About

If one picture is worth a thousand words, one day at a good electronics trade show or convention can be an education in itself. Where else can you see the products of hundreds of vendors, all having to do with what's new in electronics? Where else can you meet the men and women working in your chosen field who know what is happening and are "up" on the latest trends in products, services, and employment? Where else can you acquire a suitcase full of descriptive literature, accompanied by names and addresses of companies and people that you can contact? And where else can you be seen by individuals you need to impress, if not right now, certainly in the months to come? For an electronics activist, a visit to such a convention is time well spent in the pursuit of electronics beyond the classroom.

Trade shows have been around for decades but have really come into their own only in the last 35 years. Whether they are small or large, last for a day or a week, are held in a tiny hotel or a giant convention center, or are highly specialized or quite general, they usually consist of two elements: (1) the vendor display area and (2) the seminars and technical sessions. If you are not a member of the organization sponsoring the show, you may be restricted to the vendor display section. That's all right; that is where the sights and sounds of electronics activity are to be found.

You, as a student, are very welcome to go to such shows and conventions. Indeed, many organizations specifically encourage student attendance by reducing or eliminating the nominal admission charge.

How do you find out about such shows? There are many ways. Read trade magazines, where ads for such shows are placed. Contact your local convention center with regard to what's coming up. Stop by electronics stores and ask about trade show information. Many such establishments will give you free tickets. You can also consult the business section of your daily newspaper, ask your electronics instructor, or probe around at work. If you let enough people know what you are looking for, you'll find out when the trade shows are coming to town and where they will be held.

A final note: When you attend such events, dress and act professionally. Treat every person you meet as a potential employer. Practice talking with them: Ask relevant questions, and explain and sell yourself. Keep in mind that you're out to impress as well as to be impressed.

Summer Vacation: A Time of Opportunity

Summer vacation is what every student from grade school to grad school looks forward to with excitement and relief. It is a time to get as far away from school—and school subjects—as possible. No more electronics textbooks, no more labs, no more lectures, no more electronics, period, at least for the next 3 months. It's time to have fun, enjoy life, and get away from learning. Right? Well, maybe—and maybe not.

Sure, you deserve some time off. We all need a vacation, if for no other reason than to recharge our internal batteries and allow us to be more enthusiastic and productive when we return to the main tasks at hand—in your case, learning electronics.

And yet, summer vacation can be a vulnerable time, one fraught with lost opportunities. Three months away from electronics, where you do no reading, no building and repairing of electronic devices, no visiting of electronics industries, and no reviewing of previously learned material—in other words, where you forget about electronics altogether—is no good. You want to make sure that you enter the fall semester with *at least as much* knowledge as you left with in the spring. You can't afford to forget what you have learned about ac/dc circuits, solid-state devices, digital electronics, communications, industrial controls, medical electronics, and the like. At the very least, summer vacation needs to be a time for review, synthesis, and reflection on what you have already learned.

But it can also be considerably more than that. If you do nothing in electronics, not only will you forget much of what you know, more importantly, you will miss the opportunity to jump *ahead*. Summer vacation, in addition to providing the rest and relaxation you deserve, represents an opportunity to leap forward, become well prepared for the coming semester, *and* do the things in electronics beyond the classroom that you might not have the opportunity to do at any other time.

The key is to stay involved with electronics, no matter what that may entail. If you have a chance to work in the field, that's probably the most productive thing you can do. If not, you can visit local industries (or those out of town while on an extended vacation); that is, you can take that time to go on your own personal field trip. Many companies have scheduled tours for the public. Call a few and get out to see them. Of course, this is the ideal time to immerse yourself in the project building we discussed earlier. And if there are trade shows and conventions in your area, you'll have all day to attend them.

Finally, you might be able to get a head start on the classes you will actually be taking in the fall. You could purchase textbooks now, assuming you know which ones they are. Reading ahead in such books can never hurt. But there may be a better way to accomplish much the same thing without the "drag" of consuming a textbook while lounging around the pool or at the beach. Summer is a time for light reading. Therefore, try purchasing an inexpensive, general trade (or consumer) book on the same subjects you will be studying in the fall. Radio Shack, Hayden, Tab, and Sams publish dozens of such paperback books on every conceivable subject, from *How to Install Car Speakers* to *Understanding Solid-State Electronics*.

The point of all this, of course, is to make the summer a time of opportunity in your electronics education. No matter what else you do, what other obligations you may have, it's a great time to become a full-time "electronics activist."

The Information Interview: To Know and Be Known

A time-honored way of getting out into industry and seeing what's going on while still a student is with the **information interview.** Through it, you visit a company, talk with a technician, supervisor, manager, or even company president, and learn all you can about the company and its people. The interview usually lasts from a half-hour to an hour and is often accompanied by a tour of the facilities.

Such an interview, if it is to be effective, should be a two-way opportunity. Although your purpose is to see, it is also to be seen; while it is to hear, it is also to be heard; while it is to be impressed, it is also to impress;

and while it is to find out, it is also to be found out. In other words, while you want to learn all you can about the company and its employees, you want to let them know about you, too. You're on a scouting trip, you're looking for a potential employer, and you're out to make contacts—and the information interview is an excellent way to begin.

How do you set up such an interview? Finding a company to visit is not difficult. Begin by asking your electronics instructor for the names of individuals on the department's industry advisory council. These folks have already made a commitment to communicate with and help students. They will surely be most accommodating to you. Also, whenever a guest speaker from industry comes to talk with your class, follow up by asking him or her for an information-interview opportunity. If you have visited a trade show or convention lately, look through the literature you acquired for company names and addresses. Former students working in the field are another excellent contact. You can also call a company's public relations department directly, tell them what you want to do, and ask them for help. And, of course, there are friends and relatives working in the field who might be able to set something up for you.

When you go for the interview, be on time (if not early), dress appropriately, be extremely courteous to all individuals you encounter, and try to remember names. Go with a set of prepared questions concerning items you want to learn about. Listen carefully to what your interviewee says. Try to keep the discussion on track; don't get off into sports, the weather, or politics. And if you are taken on a tour, observe everything around you—not just the equipment and products, but the people. Do the individuals you meet appear happy in what they are doing? Does this seem like a good place to work?

When you leave, don't leave empty-handed. Be sure you get any company brochures and the business cards of the people you meet. (You might also leave your business card, if you have one.) When you get back home, quickly set up a file for the company you have just visited.

Furthermore, when you reach home, set your follow-up procedures in motion. Immediately write a short thank-you letter to the person you interviewed. You might tell the interviewee how impressed you were with the company and the individuals you met (if this is true). Ask him or her if the company hires entry-level electronics technicians right out of school and, if so, if you can send a résumé for them to keep on file. This is a keep-in-touch as well as a follow-up letter. When it's time for you to search for that first full-time job in electronics, the file you have started now, with your information interviews, will be indispensable.

Summary

In Chapter 7, we began by looking at project building, a confidence booster. We saw how electronic kits, magazines, and books provide ways to gain hands-on experience. We explored electronics clubs and organizations. And we examined electronics certification, with emphasis on the C.E.T. and CET examinations. Then we looked at how best to use your summer vacation to stay involved in electronics. Finally, it was on to exploring the information interview, a great way to see, and be seen by, a prospective employer.

Review Questions

1. _____ _____, where materials and assembly instructions are supplied, range in complexity from a simple continuity checker to a microprocessor-controlled robot.

2. Three popular electronics magazines are _____ _____, _____, and _____.

3. Project _____, where circuit plans are presented, represent an excellent approach for project building.

4. Joining an electronics club allows you to_____ for job opportunities.

5. To get started in amateur radio, you will need a _____ and a _____.

6. Though licensing in electronics may be fading, _____ _____ is growing.

7. CET stands for _____ _____ _____.

8. A good electronics _____ _____ or _____ can be an education in itself.

9. _____ _____, in addition to providing rest and relaxation, represents an opportunity to leap forward, doing things in electronics you might not normally have time to do.

10. A time-honored way of getting out into industry and seeing what's going on while still a student is with the _____ _____.

Individual and Group Activities

1. Conduct an information interview. First, identify an individual you would like to interview at a company you want to visit. Prepare a questionnaire. Conduct the interview and tour the facilities. Write a follow-up thank-you letter, and start your information-interview file.

2. Form an electronics club or association in your department. Decide, first, what your purpose and objectives will be. Determine if it would be best to affiliate with a national organization or remain independent. Gather around you a core group that will act to get the club started. Publicize your first meeting. At that meeting, solicit input from all in attendance, but bring to the meeting a well-thought-out plan of action. Take it from there.

3. Identify any licensed hams in the department, students as well as instructors. If there are any, explore the possibility of forming an amateur radio club, with the station located at the school. Contact the ARRL for information and advice.

4. Build an electronics project. Obtain catalogs from the various companies listed in Appendix 1. Select a moderately priced, easy-to-build, useful kit to get started with. Build the project and get it working. Display it in class. Ask your instructor if you can build demonstration projects to be used as teaching aids in the class.

5. Individually or as a class, attend an electronics trade show or convention. Gather company literature and the names of contact people. When you return to class, exchange contact names and impressions of the show.

6. Develop a summer vacation learning plan that, while providing for rest and recreation, will also allow you to review the previous semester's work, get a head start on next semester, and do fun things with electronics.

7. With the permission of your instructors, arrange to visit local high schools to explain your electronics program. Students only a few years out of high school themselves can often relate well to high-school audiences. Bring examples of projects and maybe a class-made video of work done in the department.

Issues for Class Discussion

1. In addition to those ways discussed in this chapter, explore other ways to become an electronics activist and take electronics beyond the classroom. What about product-repair days at school, where students are encouraged to bring products in to repair? How about forming a department library, with books, electronics magazines, and audiovisual aids? What about building a display cabinet to show off student-built projects?

2. Discuss the issues surrounding licensing and certification. Why was the legal requirement for an FCC first class general radio-telephone license dropped? Is the license still worth getting? Why or why not? Why would certification be valuable to have? Do employers really look for it?

3. Discuss the pros and cons of going to summer school. Is it best to get away from school altogether, or can you really get ahead by going to summer school? If you do go, should you take electronics courses or concentrate on general-education requirements?

Information Technology (IT) Certification:
A New World of Credentialing

Objectives

In this chapter you will learn:

- Why IT certification is so popular.
- How to identify IT certifications most likely to interest electronics technicians.
- How IT certifications compare to college degrees.
- What IT training resources are available.
- What determines IT training costs.
- How extensive and available IT testing is.
- Why A+ certification is so critical in today's job market.
- What A+ certification training consists of.
- About A+ certification testing.

- Why Network+ is the logical certification to follow A+ certification.
- What i-Net+ certification is about.

The numbers are impressive, even if the wide range dampens their accuracy, somewhat. Depending on your source of data, in the United States the need for additional IT workers is anywhere from 340,000 to 1.6 million. And with the demand for the appropriately skilled far exceeding supply, from 220,000 to 843,000 of these positions are likely to go unfilled. Meanwhile, more than 130,000 new high-tech jobs will be created yearly through 2008, according to Duncan Anderson, president of Global Knowledge, Inc., the world's largest independent information technology training company. (Computer support specialist, as a job category, will increase by 102 percent between now and 2008, making it the second fastest growing occupation, exceeded only by computer engineer at 108 percent, according to the Bureau of Labor Statistics.) With a total U.S. IT workforce of 10 million, this shortfall in tech jobs means one job in every dozen goes vacant.

Why is this so? Is there little interest in the field, for working in information technology and telecommunications industries? Not at all, if the insatiable demand for IT certification is any indicator of career aspirations. As we discover in the pages to come, a new world of competency-based certification has come into its own, the purpose of which is to provide a recognized credential, representing a standard of knowledge employers in information industries throughout the world can readily accept.

Currently, over 350 IT certifications exist, ranging from A+ certification to Cisco Certified Internetworking Expert (CCIE). That's up from just one, Certified Novell Engineer (CNE), in 1989. As of early 2001, over 2 million IT certifications had been awarded worldwide. According to Clifford Adelman, in his excellent booklet, *A Parallel Postsecondary Universe: The Certification System in Information Technology,* published by the Office of Educational Research and Improvement, U.S. Department of Education (887-433-7827), "There are general field certifications, sub-field specialty certifications, requirements and electives, hierarchies of credentials analogous to the associate's, bachelor's, and master's levels, and the equivalent of academic societies with their annual conventions matched to each."

IT certification, it's big and getting bigger. Electronics technicians need to know what's going on, not only because they will be interacting with individuals so certified, but because they are themselves in an excellent position to seek and obtain many such certifications—at least those with a hardware base, such as A+ and Network+ certifications.

In this chapter we survey the world of IT certification, looking at the relevant types, suggested study methods, test-taking suggestions, the exams themselves, and costs. We then zero in on the three key certifications, the ones most likely to be of interest to electronics technicians: A+, Network+, and i-Net+ certifications, with emphasis on the first, A+ certification.

Why All the Excitement?

Recently, a Dallas-based research company placed the following ad, for a *Data Center Operator,* on the Internet:

- Location: TX, Dallas
- Start Date: ASAP
- Employment Type: full-time, regular
- Pay Range: $35,000–$45,000
- Database: SQL Server
- Work in our Dallas Data Center of over 100 servers ensuring 24X7 up time and eliminating redundancy. The successful candidate will have an associates degree (or A+ certification with 2–3 years experience in a computer support field). Expertise in Compaq and Dell servers, NT operating system, and networking equipment such as Cisco & 3Com routers, hubs and switches, a must. SQL knowledge a plus.
- Sense of humor required!

Did you notice the reference to associate degree or A+ certification? With both, of course, your chances of landing this relatively low-end technical position in the IT world are even better.

Here's how another company ad for a more advanced position characterized its needs and offerings:

> We are seeking a Network Administrator to join our dynamic IS&T team. Responsibilities include maintaining a 200-node LAN, maintaining NT, Web and Mail servers, 4 Novell servers, and managing our help desk function which provides user support for networking and software issues.

Potential candidates should have 4–5 years network administrative experience, 2–3 years Novell network experience, and 1–2 years of Windows NT experience. Candidates must have experience with hardware and software installation, support, configuration, and maintenance, and should have a working knowledge of hubs, bridges and routers, and security management. A+ and either advanced Microsoft or Cisco certifications a plus. Salary range: $80,000 to $90,000, with an excellent benefits package.

Obviously, not an entry-level position for a recent electronics technician graduate. Nonetheless, something to strive for. With additional course work in computer science and attainment of the recommended certifications, a hardware-oriented technician would have a serious shot at such a position.

So, the jobs are there, the need is strong. It's an exciting time to be in technology, particularly information and electronics technology.

But, must one be certified to land a job in the IT world, to keep one? In some situations, yes. But in most cases industry chooses to *reward* rather than *require* certification. According to Adelman:

Industry certifications, whether in information technology or other fields, replace neither experience nor degrees. Nor do they pretend to represent an assessment of the full range and depth of knowledge, skills, or potential contribution to organizational productivity. Instead, we might say, the certification serves to *augment* experience and traditional credentials.

Such rewards, however, take tangible form as evidenced in an industry-wide, post-certification premium of 4 to 14 percent in the combination of base pay plus bonuses for Microsoft and Novell certifications, according to Adelman. Actually, the average salary reported by an MSCE (Microsoft Certified Systems Engineer) with at least 6 months experience with Cisco products is $71,800.

What are the major IT certifications? Which ones should electronics technicians be keeping an eye on? Here is a partial list, as reflected in information technology job advertisements from April 1998 through April 1999:

MCSE (Microsoft Certified Systems Engineer)

ODBA (Oracle Database Administrator)

CNA (Certified Novell Administrator)

CNE (Certified Novell Engineer)

Cisco Certifications

Other Certifications

Among "other" certifications mentioned were:

ASE (Accredited Systems Engineer, Compaq)

PSS (Certified Professional Server Specialist, IBM)

RHCE (Red Hat Certified Engineer)

CLP (Certified Lotus Professional Application Developer)

MCSD (Microsoft Certified Solution Developer)

MINS (Master of Network Science, 3Com)

A+ (PC support service)

Of course, these are but the more popular ones. There are certifications in programming, network management, training, and telecommunications. Certifications come from specific vendors as well as generic certifications from industry associations. As we said earlier, over 350 certifications and growing.

But how do these IT certifications actually function? Are they like degrees? Are they similar to licenses? As Adelman explains:

> Some certifications function like degrees. Other certifications function like licenses in that they require renewal. For example, after one receives the mantel of Certified Internet Webmaster one must work at the role for two years and complete 30 hours of continuing education to qualify for recertification. In all the Microsoft Certified Professional programs, as exams are retired and replaced, one is required to take the new assessments to remain certified. Novell adds examinations as the state of knowledge changes, and in unmistakably uncompromising language: "All CNE certified individuals are required to pass the exam for either course 529 [Netware5 Update] or 570 [Netware5 Advanced Administration] by August 31, 2000. Failure to do so will result in loss of CNE status."

While earning certifications is not the same as earning college degrees, to a certain extent the two can be compared. The Council on Computing Certification is developing a hierarchy of credentials. There is to be a Level I, II, and III. In a way, these levels will be analogous to the associate, bachelor's, and master's degrees—at least in the context of one's major field of study. However, having your A+ certification, for example, isn't the same as having an associate degree in electronics technology, nor will it ever be. The best approach? Consider getting both.

Do You Have What it Takes?

So, how do you go about it, how do you gain knowledge and skills, and prepare for the various certification examinations? According to a Microsoft 1997 salary survey of sample certification holders, 98 percent indicated self-study as a preparation method, with 91 percent using books. A similar study revealed that 43 percent of 6000 certification candidates indicated self-study as their primary preparation method. Interestingly, college course work was last in this survey, though other classroom training (from vendors themselves) was ranked much higher.

Speaking of vendors, just who are the providers of training materials and programs? Essentially, there are *primary vendors,* such as Microsoft, Sun, Cisco, and Novell. There are *training partners,* such as Global Knowledge, Azlan, and New Horizons. And various *colleges,* notably community colleges.

As an example, Microsoft Training & Services offers you, for a fee, the following training resources when seeking to pursue IT certification:

- Microsoft Official Curriculum (MOC)
- Microsoft PressR
- MCP Approved Study Guides
- Microsoft Certified Technical Education Centers (Microsoft CTECs)
- Microsoft Authorized Academic Training Program (AATP)
- Microsoft Technical Information Network (TechNet)
- Microsoft Developer Network (MSDN)
- Seminar Online

And if you are interested in learning to install, maintain, and troubleshoot CiscoR routers, SmartCertify Direct offers CBT (Computer-Based Training) courseware covering the following topics:

- Fundamentals of Cisco Router Configuration
- Installation and Maintenance Procedures
- The 2500 and 3600 Series

- Installing and Maintaining the 4000 Series
- Installing and Maintaining the 7000 Series
- Installing and Maintaining the 7500 Series
- The 7200 Series and 7000 Family IPs
- Installing and Maintaining the 12000 Series I
- Installing and Maintaining the 12000 Series II
- Router and Route Switch Modules

The cost for all of this training? It depends. Is the provider a public community college or a for-profit vendor? Is coursework delivered on-line or live, from an instructor in a seminar setting? Does the course require extensive hands-on laboratory activity or is the format mostly lecture/demonstration? These and other factors will affect your cost for training. A recent survey revealed costs as low as $600 for A+ certification training to a high, from one private vendor, of $20,330 for Certified Internet Webmaster training.

Training is one thing, but who actually does the testing? Three testing companies play major roles: Prometric, CatGlobal, and Virtual University Enterprises (VUE). They are in the business of testing competence from PowerPoint to C++ to the elements of the Citrix CCA and those of the major industry certifications (MCSE, ODBA, etc.).

If you think only three testing companies means meager accessibility for test takers, you're in for a pleasant surprise. Prometric operates about 2500 testing centers in 140 countries. VUE has close to 1500 locations, including 20 in Mexico, 28 in Russia, 23 in Brazil, 19 in South Africa, and 50 in China. And CatGlobal offers wholly on-line computer-based testing from servers in 16 countries.

Of course, the tests themselves are not free, there is a fee. A given certification may require as many as nine examinations, fees for which range from $50 to $250. And for performance assessments with hands-on laboratories and simulations, you're talking thousands of dollars. For instance, if one completes assessments on any three tracks of the Cisco Certified Internetworking Expert program, each of which requires both a paper examination and a laboratory performance, the total examination fee could come to $3300, according to Adelman's report.

What about those exams? What are they like? How tough are they?

First, they all have cut-scores. This is similar to what occurs in licensure for law, medicine, architecture, and nursing. It's a "no pass/no play" game, if you can call such serious testing a game. Of course, as with most licensure exams, you may retake the certification tests, with an appropriate wait between test administrations.

Exam formats range from 45 to 120 minute restricted response (multiple choice, identification, one *best* correct answer, 2/3/4 correct answers), to restricted-response adaptive modes. The latter are interesting. In the adaptive mode, the testing process is stopped at a point in the adaptive curve at which a passing score could be predicted at a 95 percent confidence level. There are also constructed response and essay exams.

The most demanding certification, as we mentioned earlier, is thought to be the Cisco Certified Internetworking Expert (CCIE). To become certified, you must take a 2-hour written examination and a 2-*day* lab exam that "pits the candidate against difficult build, break, and restore scenarios." It ain't easy.

Yet, there's plenty of help available for those en route to a certification goal. As Adelman points out:

> Unlike the competitive environment for access to elite undergraduate institutions or to "top" law and medical schools, the spirit of the IT certification industry involves cooperation and sharing. The most revealing indication of this phenomenon is the "braindump," a web site at which prospective examinees seek—and others provide—sample questions and problems that will help them prepare.

Still, help aside, certification examinations are not the type you can pass by interpreting verbal clues in prompts. You must know the "stuff." You must study hard. A Microsoft 1998 salary survey revealed that those who earned the MCSE in the previous year spent an average of 216 hours preparing for their certification exams. Hard work, but, by all accounts, well worth it.

A+ Certification: The Place to Begin

Ken Freeman, downsized out of a job, no, out of a career, in banking in the early 1990s, the 40-something collector of Harley-Davidson motorcycles needed to establish a new direction. After an introductory course in electronics at LA Valley College, he found it. Two years later, Ken was graduating with an A.S. degree and heading out into the job market as a computer support technician.

Then, 2 years after that he saw that his alma mater was looking to hire a full-time Information Technology Lab Technician. "A.S. degree required, A+ certification desired," the job posting declared.

"I had the former," said Ken. "And I had taken course work in preparation for the A+ certification exam. But I just hadn't bothered to follow through on taking the test."

Three months later, after intensive review and study, Ken was ready to rectify his mistake. Then, armed with both a degree and A+ certification, he applied for the lab tech job.

"Now I work where I recently went to school," Ken says. "I love it! Clearly, having my A+ certification made the difference." (See Figure 8.1.)

Ken is representative of the more than 135,000 A+ certified technicians in the United States. He, like them, has discovered that, essentially, A+ certification is becoming a requirement for entry-level technicians in the computer service industry.

"We require A+ certification before a technician can take our Warranty Authorization Training Program," says Dennis O'Leary, IBM, Manager of Global Education Programs. "When we 'train up' from the A+ validated level of competence, we eliminate redundancy and save money."

Figure 8.1
A+ certificate.

Core Exam

Domain	% of Examination
1.0 Installation, Configuration & Upgrading	30%
2.0 Diagnosing and Troubleshooting	20%
3.0 Safety and Preventive	10%
4.0 Motherboard, Processors, Memory	10%
5.0 Printers	10%
6.0 Portable Systems	5%
7.0 Basic Networking	5%
8.0 Customer Satisfaction	10%

Windows/DOS Specialty Exam

Domain	% of Examination
1.0 Function, Structure, Operation and File Management	30%
2.0 Memory Management	10%
3.0 Installation, Configuration and Upgrading	25%
4.0 Troubleshooting	25%
5.0 Networks	10%

Figure 8.2
A+ certification examination objectives with weighted percentages.

"A+ certification is an important consideration for managers, technicians, and technology instructors," adds James B. Lacey, CompUSA, Director of Instructional Development. "Furthermore, this certification is a condition of employment with major resellers, and a foundation for those who wish to increase their technical knowledge in the IT industry."

The bottom line: many businesses require A+ certification as a prerequisite to hire, requiring new employees to attain the A+ certification within 90 days to continue employment. Many hardware vendors will only allow A+ certified technicians to perform warranty service. And, if you are looking for the bottomline, the average salary of an A+ certified technician working in the U.S. is reportedly $49,000 a year.

But just what is A+ certification? How did it come about?

Sponsored by CompTIA (Computing Technology Industry Association), A+ certification is the industry-wide standard for measuring benchmark level, vendor-neutral technical skills expected of a computer service technician with 6 months on-the-job experience. Tests are administered by Prometric. The program is backed by over 50 major computer hardware and software manufacturers, vendors, distributors, resellers, and publications. A+ certification came about to eliminate redundant training and reduce the associated costs of preparing service technicians for entry-level positions.

To gain A+ certification, you must pass two exams: the **Core** exam and a **Specialty** operating system exam. Basically, the Core exam deals with hardware, the Specialty exam with software (DOS/Windows).

In the Core part, you demonstrate your ability to properly install, configure, upgrade, troubleshoot, and repair microcomputer hardware. The Core tests general skill rather than knowledge of specific manufacturer systems or components. As shown in Figure 8.2, it consists of eight domains. Note the weighted percentages of importance.

In the Specialty operating system exam you demonstrate competency in installing, configuring, upgrading, troubleshooting, and repairing microcomputer systems. Figure 8.2 shows the five domains that make up this portion of the test, along with their weighted percentages of importance.

A+ Certification: The Content/The Test

So, what is actually covered in the various sections, Core and Specialty? What, specifically, must someone know?

While we cannot outline all the domain areas here, looking at one in detail is instructive. Here is the Domain 1.0 Installation, Configuration, and Upgrading outline, as presented on Specialized Solutions, Inc.,'s web site (www.specializedsolutions.com/A+_kit.htm):

> This domain requires the knowledge and skills to identify, install, configure, and upgrade microcomputer modules and peripherals, following established basic procedures for system assembly and disassembly of field replaceable modules. Elements include ability to identify and configure IRQ'S, DMA'S, I/O addresses, and set switches and jumpers.

Here are the individual subsection descriptions:

1.1 Identify basic terms, concepts, and functions of system modules, including how each module should work during normal operation.

1.2 Identify basic procedures for adding and removing field replaceable modules.

1.3　Identify available IRQ's, DMA's, and I/O addresses and procedures for configuring them for device installation, including identifying switch and jumper settings.

1.4　Identify common peripheral ports, associated cabling, and their connectors.

1.5　Identify proper procedures for installing and configuring IDE/EIDE devices.

1.6　Identify proper procedures for installing and configuring SCSI devices.

1.7　Identify proper procedures for installing and configuring peripheral devices.

1.8　Recognize the function and effective use of common hand tools.

1.9　Identify procedures for upgrading BIOS.

1.10　Identify hardware methods of system optimization and when to use them.

Yes, a lot of identifying. After all, this is an exam you're preparing for.

What's the examination like, and how is it administered? According to a CompTIA brochure, entitled *A+ Certification: A CompTIA Certification Program:*

> The A+ examinations are administered on a computer at a Sylvan Prometric authorized testing center. In an easy-to-use format, the examination looks very much like other multiple-choice tests. Because the desktop computer is connected into a testing network where your answers are stored, you receive final scores as soon as you finish the test. However, your answers to specific questions are securely stored and never released.
>
> There are 69 Core questions and 70 DOS/Windows questions on the two-part test. Questions appear in situation (choosing a situation or scenario commonly encountered by service technicians), identification (selecting an answer that identifies what is being shown in a diagram, flowchart, or illustration), and traditional (picking the correct answer from a list of choices) formats. It is recommended that you take both parts of the test on the same day. If not, you must pass both parts within 90 calendar days to be awarded certification.

Time to start studying. Why not earn your A+ certification along with an A.S. degree?

As we said earlier, A+ certification need not be an end in itself, far from it. In Figure 8.3, we see two possible paths from A+ certification leading

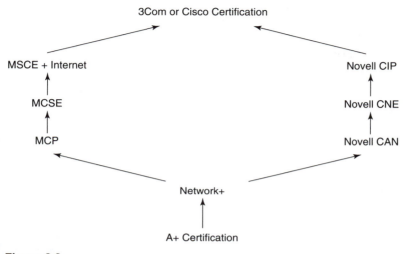

Paths to IT Success

Figure 8.3
Possible paths from A+ certification to 3Com or Cisco certifications.

onward to 3Com or Cisco certification. Regardless of which route you take, Novell or Microsoft, your immediate next stop is Network+ certification. Let's take a moment to see what it's all about.

Network+ Certification: Your Next Step

Network+ Certification is also a CompTIA testing program. It seeks to authenticate the certification of IT technicians with at least 18 to 24 months experience. Certification means you possess the knowledge needed to configure and install network components.

The Network+ examination is broken into two distinct parts: **Knowledge of Networking Technology** and **Knowledge of Networking Practices.** The exam consists of 65 questions and must be completed within 90 minutes.

As shown in Figure 8.4, the Knowledge of Networking Technology part is made up of nine domains. The total weight for this portion of the exam is 77 percent.

I. Knowledge of Networking Technology

Domain	Percentage of Examination
Basic Knowledge	18%
Physical layer	6%
Data Link Layer	5%
Network Layer	5%
Transport layer	5%
TCP/IP Fundamentals	16%
TCP/IP Suite: Utilities	11%
Remote Connectivity	5%
Security	6%

II. Knowledge of Networking Practices

Domain	Percentage of Examination
Implementing the Installation of the Network	6%
Maintaining and Supporting the Network	6%
Troubleshooting the Network	11%

Figure 8.4
Network+ certification examination objectives with weighted percentages.

The Knowledge of Networking Practices section consists of four domains, totaling 23 percent.

A survey conducted by the University of Maryland University College (UMUC) reveals that the average annual salary for the Network+ certified individual in 1998 was $64,900.

Another certification worth investigating is the relatively new i-Net+ certification. Also sponsored by CompTIA, its purpose, according to the organization, is to:

> Target individuals interested in demonstrating the baseline of technical knowledge that would allow them to pursue a variety of Internet-related careers. The i-Net+ exam was specifically designed to certify entry-level Internet and e-commerce technical professionals responsible for participating in the maintenance of the Internet, Intranet, and Extranet infrastructure and services as well as the development of Web-related applications.

Susan Davidson, worldwide program manager of the Intel Certification Program, had this to say about i-Net+ certification:

Areas of Knowledge

Domain	Percentage of Examination
Internet basics	10%
Internet clients	20%
Development	20%
Networking	25%
Internet security	15%
Business concepts	10%

Figure 8.5
i-Net+ certification examination objectives with weighted percentages.

CompTIA's i-Net+ certification encompasses the core technologies used in the Internet economy and represents an excellent validation of skills needed to begin building secure, reliable e-business solutions. As a result, Intel accepts the i-Net+ certification as meeting the baseline requirements for the Intel Certified Integration Specialist e-business Implementation (ICIS) track.

The i-Net+ exam tests six areas of knowledge, as shown in Figure 8.5. The weighted percentages shown are approximate as of this writing and are subject to change.

IT certification, from A+ certification onward is now an integral part of what determines qualification in technology industries. It is not a substitute for experience or degrees, as we mentioned earlier. Yet, if you are to become an electronics technician, particularly in a computer-related field, gaining such certification is well worth the effort.

Summary

In Chapter 8, we began by looking at the breadth and depth of IT certification. We saw why employers are encouraging, and in a few cases requiring, such certification of their employees. We examined training institutions and test-taking enterprises. And we looked at the examinations themselves, seeing what cut-scores and restricted-response adaptive mode testing are about.

We then took an in depth look at A+ certification, the IT certification most likely to be of interest to newly minted electronics technicians. We examined a sample A+ certification curriculum and discussed how the A+ exam is administered. We concluded with a brief look at Network+ and i-Net+ certifications.

Review Questions

1. Currently, over _____ IT certifications exist, ranging from A+ certification to Cisco Certified Internetworking Expert (CCIE).

2. Industry certifications, whether in information technology or other fields, replace neither _____ or college _____.

3. MSCE stands for _____ _____ _____ _____.

4. The Council on Computing Certification is developing a hierarchy of credentials with three levels, analogous to the _____, _____, and _____ degrees—at least in the context of one's major field of study.

5. In providing IT certification training materials and programs, there are _____ vendors, training _____, and various _____.

6. With regard to IT certification, there are three testing companies: _____, _____, and _____ _____.

7. All IT tests have a _____-score.

8. In the restricted-response mode, the testing process is stopped at a point in the adaptive curve at which a passing score could be predicted at a _____ percent confidence level.

9. The A+ certification examination consists of core and specialty parts, the former dealing with _____, the latter with _____.

10. Network+ certification means you possess the knowledge needed to _____ and _____ network components.

Individual or Group Activities

Summarize the results of each activity in a 2- or 3-page report or a 5-minute oral presentation to the class.

1. Why not survey your classmates to see how many are interested in earning an IT certification? In which certifications are they interested? Why?

2. Survey local information technology industries to find out how many encourage IT certification by their employees. Do they actually require such certification?

3. Develop a listing of local A+ certification testing sites.

4. Check the newspaper want ads to see which companies are asking for IT certification. Cut the ads out and post them in class.

5. List a half-dozen IT certifications you might be interested in and surf the Internet looking for information on each.

Issues for Class Discussion

1. Are IT certifications really like college degrees? In what ways might this be so, in what ways might they not be?

2. Discuss IT exam preparation techniques. Is self-study the way to go? Can you learn all you need to know from a book?

3. Discuss restricted-response adaptive mode testing. Could such a method work in your regular college classes?

4. Why do you suppose web sites such as "braindump" exist? Why are individuals willing to help others succeed in getting IT certified? Aren't such individuals worried about the resulting competition?

5. Discuss ways in which you and your classmates might form an A+ certification study group. How might you share resources?

CHAPTER 9

Launching Your Career as an Electronics Technician: Finding, Getting, and Keeping Your First Position

Objectives

In this chapter you will learn:

- Why just being out of school is an advantage in the job search.
- The best job search strategies.
- How the Internet is simplifying the job search.
- What you should be doing each semester to enhance your job prospects.
- How to write a simple, effective, and flawless résumé.
- The purpose of a cover letter and how to write one.
- How to fill out a complex job application.
- How to make the most of your first few weeks on the job.
- About ten "no-no's," likely to upset any employer.

Do you remember Bob in Chapter 7, the one who got the position at ABC Electronics? Recall how self-assured and confident he was and how easy it seemed for him to get the job? In truth, it wasn't all that effortless, and Bob didn't always possess such composure and poise. Such qualities took time to develop and mature. Furthermore, his getting the job at ABC Electronics was the culmination of a great deal of effort, planning, and preparation. It represented the fruition of a 2-year job-search plan that Bob put into effect the day he started his studies to become an electronics technician. For Bob, obtaining that first full-time job as an electronics technician didn't just happen; he made it happen.

Bob learned early that he needed to complete his formal, 2-year technical education having in his possession two types of skills: (1) those necessary to *do* the job and (2) those necessary to *get* the job. It wasn't enough to know electronics; he also had to acquire the strategies and know-how essential to getting a job in electronics.

In this chapter, we show you how to develop and hone those strategies, too. We start with something that might surprise you: an explanation as to why you, a *student* fresh out of school, may actually have a better chance of getting a job as an electronics technician than a more seasoned competitor. From there we go on to the three big steps: *finding, getting,* and *keeping* that position you have been studying and planning for all along.

Before we begin, though, let's consider a word of caution. In just one chapter, we cannot hope to cover such a vast subject in any depth. Indeed, entire books have been written not only on this topic, but also on subsections of it, such as "How to Write a Résumé" or "How to Handle the Job Interview." All we can do here is alert you to what needs to be done and get you started.

Throughout, keep in mind that your transition from school to work can be exciting, though at times jolting. As a student, you did the learning and you paid someone or some organization for the privilege. As an electronics technician, you are now going to give what you know and what you can do, and you are the one who is going to be paid. Believe me, it's quite a "high" to find out that someone, some company, is actually prepared to pay you, supply your medical and dental care, put your kids through college, etc., all because you have something to offer. And yet, that's how it should be, given that you have studied and worked so hard. You deserve to succeed—to get that job. However, don't forget what Bob found out—it won't just happen; you have to go after it. So, with that in mind, let's get started.

Me: Just a Kid Out of School?

We hope this will make your day! If you're like a lot of students just completing an electronics education, you are not only a bit apprehensive about what you know and your ability to apply it in an industrial setting, you're also nervous about competing in the job market with experienced, practiced electronics technicians. Comments such as the following abound:

> "I know a guy who has been working at Hughes Radar as a high-definition radar technician for over a decade. He wants to get out of aerospace—how can I compete with him?"
>
> "My cousin has been fixing computers for 8 years. That's 8 years in electronics that I don't have."
>
> "James has an older brother who says he's a tech, works for Sears, I think. He is looking for a new job. He's experienced; I'm not. What's that going to mean?"

Not to disparage the value of experience, but there are at least three good, documented reasons why you, a "newly minted" technician, may, in many cases, actually have the advantage.

First, many companies, especially the large and well-established ones, like to work directly with community colleges and private technical schools. This is because in doing so they know what they are getting. They know the curriculum and they know the type of graduates that are being turned out. They know pretty much what you, the student, know and don't know. As Frank Garcia, of Signetics, says: "We work directly with the local college, we're on their advisory committee, we help shape the curriculum, students work with equipment we have donated. In short, from the college—what we see is what we get."

Conversely, with an experienced tech, do you know what you're *really* getting? Does a decade immersed in the specialty of high-definition radar qualify one for work with PC networks? And does the résumé actually illuminate the true experience of an individual? Dean Hyde of CLS Industries related the story of a fellow who applied for work as a components test technician. His résumé indicated 3 years experience in this subfield. According to Dean: "As it turned out, he worked as a parts and components dispenser from the supply room. Nothing wrong with that, except it's certainly not technician-level work."

Second, being right out of school, you're ready for more school; you are "learning ready." In many companies, upon being hired you'll jump right into a 2-week to 6-month training program. Chances are, as a recent graduate, you're ready for that sort of thing far more than "old-timers" who may feel it is beneath their dignity to go back to school or, worse, just wouldn't do well if they did. You, on the other hand, can hit the ground running—a key advantage.

Third, many companies want to train someone from the outset on company procedures and equipment. They want to mold you "their way" and bring you along in "their image." Rick Cunningham of Advanced Electronics said it well: "We like to hire beginning techs because they will then grow with us, they won't get bored with the job, and they will see a future for themselves and stay here."

Also, often such companies are looking specifically for entry-level techs because the company's investment in such individuals isn't as great. They can pay you less while checking you out. If you don't work out, their costs are less than if they went with a more experienced technician.

All this isn't to say that experience doesn't count or that it is of little value. On the contrary, when such experience is real, relevant, and recent, it can be invaluable. Nonetheless, as the preceding comments indicate, being fresh out of school is not only far from being a detriment, in many cases it is an asset. Use that asset to your advantage.

Finding Your First Position as an Electronics Technician: The Job Search

Your first step in launching an electronics career is to find companies that hire electronics technicians. In doing so, you need to know what type of job (full-time, part-time, etc.) is best for you, what kind of companies hire electronics technicians, where such companies are located, and what job-search methods are most effective. In the process, you will want to develop a two-year job-search strategy that, ideally, begins with your first day at school.

If you're about to graduate—that is, complete your certificate or degree—chances are you will be looking for a full-time, entry-level position as an electronics technician. After all, this is what you have been studying for, it's been your career objective all along, and you need to get on with your life and earn a living wage. For the majority of electronics technician students, full-time technician work will be the most sought-after position.

There are alternatives, however, that should be touched upon. Some companies may, at the present time, be hiring only part-time technicians (thus eliminating the need to pay benefits). If you can afford to work part-time, this may be your foot in the door. Another way to get started is by accepting a *bridging position,* a nonelectronics job that gets you on board. You then keep your eyes and ears open for a lateral move into electronics. Finally, there is the possibility of going into business for yourself. Though enticing to some, such a momentous step is perhaps best postponed until you have gained more experience in electronics and the world of work in general.

These last alternatives aside, let's get back to full-time electronics work. What type of companies actually hire electronics technicians? First, of course, there are the *electronics companies* that produce the products and services of the electronics age. Advanced Electronics of Gardena, California, is typical. They sell, install, and repair Motorola communications equipment, employing 25 full-time electronics technicians. Also, in this category are the *vendors* of electronic devices. For example, Toshiba-America of Tustin, California, manufactures and distributes medical electronics equipment. Not only do they employ electronics technicians at their plant, but they send those individuals out to customer sites (usually hospitals) to troubleshoot and install the equipment.

There are, in addition, many *nonelectronics companies* that employ electronics technicians of all types. Hospitals are a good example. Most hospitals of any size have medical electronics technicians on duty to deal with emergency equipment failure and routine problems. If more extensive work is required, the distributor, or vendor (Toshiba-America, for example), is called in. It is important to keep these nonelectronics companies in mind while engaging in your job search. Electronics is so pervasive that it's hard to find any enterprise immune to it. As an electronics technician, you're going to wind up where electronics is—and that's everywhere.

Speaking of everywhere, what about job location? Are the electronics jobs concentrated in particular geographic areas? Yes, but then so is the competition for those jobs. There are certainly more electronics firms in Los Angeles, California, than in Boise, Idaho. But there are technician jobs in Boise, and if that's where you want to live and work, begin your job search there. Of course, you may not find what you are looking for and will then need to expand your geographical horizons. The point is this: Don't write off any town, city, country, or state just because you think the high-tech jobs aren't located there. There are all sorts of reasons why that just isn't so.

It's time now to turn to specific job-search procedures. Just what are the methods used to find technician positions and which ones are most effective? Table 9.1 can give us a clue. In it we see listed a dozen job-search

TABLE 9.1
Commonly used job search methods.

Percent of total job-seekers using the method	Method	Effectiveness rate (%)*
66.0	Applied directly to employer	47.7
50.8	Asked friends about jobs where they work	22.1
41.8	Asked friends about jobs elsewhere	11.9
28.4	Asked relatives about jobs where they work	19.3
27.3	Asked relatives about jobs elsewhere	7.4
45.9	Answered local newspaper ads	23.9
21.0	Private employment agency	24.2
12.5	School placement office	21.4
15.3	Civil Service test	12.5
10.4	Asked teacher or professor	12.1
1.6	Placed ad in local newspaper	12.9
6.0	Union hiring hall	22.2

* A percentage obtained by dividing the number of job-seekers who actually found work using the method, by the total number of job-seekers who tried to use that method, whether successfully or not.
Source: Tips for Finding the Right Job. U.S. Department of Labor, Employment and Training Administration, 1991.

methods, the percentage of job-seekers using each method, and the effectiveness rate. Keep in mind that the information covers all job categories, not just that of electronics technician.

According to the data, 66.0 percent of all job-seekers apply directly to an employer, and when they do, nearly half, 47.7 percent, find employment. Clearly, direct employer contact is effective. So, too, is answering local newspaper ads: 23.9 percent, or close to one out of four, job seekers find a job in this manner. Asking friends about jobs where they work is also a successful, time-honored approach, 22.1 percent finding that this method hits the mark. At the other extreme, asking a teacher or professor, or taking a civil service test, are, in general, ineffective. However, with regard to asking your instructor, keep in mind that the success rate in vocational programs is bound to be higher, where finding students a job is often an integral part of the instructor's responsibilities.

Though helpful, the list in Table 9.1 is not complete. A number of job-search approaches for the would-be electronics technician are not even mentioned. Here, then, are a half-dozen key job-search methods that electronics students who want to become full-time electronics technicians should seriously consider:

1. *Networking with friends and relatives.* Everyone you meet—relatives, friends, former graduates, or the person behind the counter at the local electronics store—needs to know that you are looking for a position as an electronics technician. This is not the time to be shy or timid—get the word out. Let people know what you want and how they can get in touch with you if something comes up.

2. *Newspaper want ads.* Check the newspapers every day, particularly the Sunday editions. Though most ads are nondescriptive, they will at least give you the names of companies and individuals to contact now or later on.

3. *School placement offices and public and private employment services.* Check your electronics department's files and those of the general placement office. Let teachers know right away what type of work you're looking for. Don't forget the public (free) employment agency. This place is not just for picking up unemployment checks. When dealing with private employment agencies, remember there is a fee, paid by you or the employer. Check things out carefully before signing anything.

4. *Job fairs.* When employers come to your school to set up a booth, hand out literature, and create their own applications list, approach them with a smile and an "I know what I am looking for" attitude. Shake hands, introduce yourself, and ask for an application and a follow-up interview. If they will accept it, present them with a résumé. Be sure to leave the booth with company brochures. Follow up with a phone call or letter of inquiry.

5. *Professional clubs or organizations.* To get a list of organizations, consult the *Encyclopedia of Associations* found in your LRC. It is an annual multivolume publication listing thousands of trade associations, professional societies, labor unions, and fraternal organizations that could be of assistance.

6. *Cold calls and walk-ins.* A cold call takes some determination but can be extremely effective. Find a list of employers in the

yellow pages of your telephone book and from the local chamber of commerce. Your objective, whether on the phone or in person, is to get an application on file. Most companies that you contact will not be hiring at the moment. But now that you are in their files, you can follow up with an inquiry.

7. And, of course, there is *the Internet.* Used by both job-seekers and companies looking for employees, the Internet is a must job-search "engine." As Cynthia B. Leshin says in her book, *Internet Investigations in Electronics* (Prentice Hall, Publisher), "The question now is no longer whether the Internet should be used to find a job or an employee, but rather, how to use it."

Cynthia has come up with a great seven-step strategy to Internet job searching. Here is what she recommends you do:

Step 1. Research companies that you are interested in by finding and exploring their Web pages.

Step 2. Explore job resources and employment opportunities available on the Internet.

Step 3. Learn about electronic résumés.

Step 4. Visit on-line sites for job seekers.

Step 5. Create an on-line résumé to showcase your talents.

Step 6. Use the Internet to give yourself and your résumé maximum visibility.

Step 7. Use the Internet to learn as much as possible about a prospective company before going for a job interview.

Clearly, on-line job searching is here. According to the investment firm of Thomas Weisel Partners, the number of on-line job seekers is expected to rise from about 1.5 million in 1999 to as many as 10 milliion in 2003. And it isn't just nerdy computer types who are using the services of on-line job finders such as Monster.com and HotJobs.com. According to Bernard Hodes, president of CareerMosaic, "While computer and engineering jobs are still widely sought online, the Web is increasingly going mainstream."

With online job search sites, you can both search for a job and be sought after. But before clicking your way through job-hunting cyber-space, remember this, according to Erin Arvedlund, a staff reporter for TheStreet.com, "Over 90 percent of resumes and applications submitted online never get an acknowledgment or response. Also, job seekers out-number positions by a margin of five to one."

FIRST SEMESTER

Goals: To let others know your long-range plans and career objectives and to learn about electronics subfields.

Suggested Activities:

1. Let everyone know your major, school, projected date of graduation, and career plans.
2. Take field trips, and explore electronics industries.
3. Listen to guest speakers from industry.
4. Attend conventions and trade shows.

SECOND SEMESTER

Goals: To join organizations that will expand your knowledge and contacts and to get to know working technicians.

Suggested Activities:

1. Start a company-employer file.
2. Interview working technicians.
3. Continue with field trips, conventions, and trade shows.
4. Join or help form an electronics club.

THIRD SEMESTER

Goal: To become more focused on specific career objectives.

Suggested Activities:

1. Conduct information interviews.
2. Attend job fairs to gather information.
3. Seek intern, co-op, and part-time work in electronics.
4. Practice interview techniques.
5. Expand after-school electronics activities.

FOURTH SEMESTER

Goal: To begin an active job search.

Suggested Activities:

1. Use the Internet to find a job.
2. Consult department and school placement offices.
3. Consult public and private employment agencies.
4. Attend job fairs and set up employment interviews.
5. Write your résumé.
6. Consult newspapers and trade magazines for employment openings.
7. Do cold calls and walk-ins with various companies likely to hire technicians.
8. Let everyone know you are looking for a position *now*.
9. Sit for actual job interviews.

Figure 9.1
Job-search strategy.

Here, according to Hunt-Scanlon Advisors, is a listing of on-line job search firms most frequently used by company recruiters:

- monster.com   (59 percent)
- CareerMosaic   (37 percent)
- Headhunter.net   (26 percent)
- America's Job Bank   (26 percent)
- JOBTRAK   (24 percent)

While exploring these services is well worth the effort, don't limit yourself to online job sites to find that perfect job, traditional methods should not be forsaken.

How many of these job-search methods should you use? You may need all of them and then some. As has been said, finding a job is a full-time job in itself.

Maybe full-time is a bit exaggerated. But a well-thought-out, long-term approach to the job-search process certainly is not. Figure 9.1 shows a 2-year, 4-semester job-search strategy that is worth considering. It's flexible; if you're on a shorter schedule, modify the plan accordingly. And the borders between semesters aren't fixed, so you may want to start interviewing technicians in your first semester rather than waiting until the second or third. Basically, the whole point of the strategy is simply to have one. It's never too early to start thinking and planning for that full-time job as an electronics technician. In the time you are in school, whether it be 6 months or 2 years, there are things to be done within each stage, or semester: data to be gathered, field trips to be taken, conventions to visit, job fairs to check out, and, when it comes time to actually get the job, résumés to write, applications to complete, and interviews to take. Throughout, a file should be kept on all your activities: materials you have gathered, people you have met, and companies you have visited. Then, by the time you're ready to kick your job-search operation into high gear, a few months before graduation, you'll know what you want and have the resources necessary to maximize your options.

Getting Your First Position as an Electronics Technician: Being the Chosen One

Until now we have dealt only with the first step toward full-time employment as an electronics technician—the job search. Once you have identified an attractive company that you hope is hiring, it is time to move on to the second, most critical step, that of actually getting, or securing, the job. Here, you must do everything possible to become the selected individual from what will surely be a deluge of candidates. To ensure that you are the chosen one (or at least among the select few if more than one position is open), you must pay close attention to four critical *job-securing factors.* You will have to (1) write a flawless *résumé* and *cover letter;* (2) methodically complete the *job application;* (3) "ace" any employee *test* that is given; and (4) not only survive, but exploit, the *job interview.* With attention to detail and some practice, you can easily do all four. Let's see how.

Your **résumé** is your personal inventory and formal introduction to a potential employer. It is required when applying for a technical position. The résumé's purpose is to get you an interview, no more, no less. It should stress what you can offer, not what you want; it must state accomplishments, not just descriptions. As a rule, résumés are not read, but skimmed; therefore, they should be kept short, preferably to one page. Keep in mind that the résumé tells an employer two things about you: what you are and who you are. Put another way, it's not just what you say but how you say it. The presentation of the résumé (and cover letter, to be discussed shortly), its neatness, the absence of spelling and grammatical errors, and its concise writing style all are as important as the information contained within.

Again, remember Bob in Chapter 7? We have his chronological résumé in Figure 9.2. Let's analyze each section in an attempt to identify key points:

1. Name, address, and telephone number are centered at the top of the page. Include your work phone number only if you don't mind being contacted at your place of employment.

2. Try to state your career objective in one sentence. Be general, but at the same time give it some definition. Keep the objective immediate, not long range.

3. List, in reverse chronological order (last school first), your education, but limit it to three schools. The format is: degree, followed by your major, name of school, location, and year of graduation. If you haven't graduated yet, state the projected date.

4. List, in reverse chronological order, your work experience. Give job title, whether work was part- or full-time, company, location, dates of employment, and the manager or owner's name. Then, in bullet format, list three or four job duties you performed.

5. In the skills section, blow your own horn. List special skills related to a position as an electronics technician.

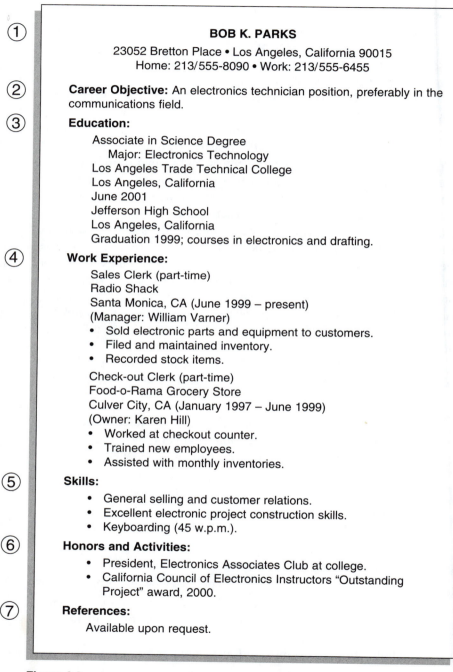

① **BOB K. PARKS**
23052 Bretton Place • Los Angeles, California 90015
Home: 213/555-8090 • Work: 213/555-6455

② **Career Objective:** An electronics technician position, preferably in the communications field.

③ **Education:**

Associate in Science Degree
   Major: Electronics Technology
Los Angeles Trade Technical College
Los Angeles, California
June 2001
Jefferson High School
Los Angeles, California
Graduation 1999; courses in electronics and drafting.

④ **Work Experience:**

Sales Clerk (part-time)
Radio Shack
Santa Monica, CA (June 1999 – present)
(Manager: William Varner)
- Sold electronic parts and equipment to customers.
- Filed and maintained inventory.
- Recorded stock items.

Check-out Clerk (part-time)
Food-o-Rama Grocery Store
Culver City, CA (January 1997 – June 1999)
(Owner: Karen Hill)
- Worked at checkout counter.
- Trained new employees.
- Assisted with monthly inventories.

⑤ **Skills:**

- General selling and customer relations.
- Excellent electronic project construction skills.
- Keyboarding (45 w.p.m.).

⑥ **Honors and Activities:**

- President, Electronics Associates Club at college.
- California Council of Electronics Instructors "Outstanding Project" award, 2000.

⑦ **References:**

Available upon request.

Figure 9.2
A résumé.

198

6. With honors and activities, you again have an opportunity to list anything that amplifies your personality and your preparation for the job of electronics technician.

7. Never supply references with the résumé. Instead, make them available upon request. (Contact three or four individuals ahead of time and ask them for recommendations.)

Once it is written and edited (proofread preferably by a friend or relative), your résumé should be printed on a letter-quality printer, and a hundred copies should be run off on an offset press or top-quality photocopier.

I am sure you can't help but notice there are books, mini-courses, and seminars galore that show you how to prepare a "killer résumé," one "guaranteed" to land you a job interview. All this résumé interest exists not only because workers are moving from job to job more frequently, but because preparing and sending a résumé, thanks to computer résumé creation programs and the Internet, has never been easier. In fact, many large companies are so inundated with unsolicited résumés. they've resorted to "employing" résumé scanning software to search for key words in your document.

In preparing a twenty-first century résumé, it's best assume it will be "read" by a computer first. Here are some tips to keep in mind when preparing a computer-friendly "vanilla" resume:

• Avoid unusual type faces, underlining, and italics.
• Use 10 and 14-point type.
• Use smooth white paper, black ink, and quality printing.
• Be sure your name is on the first line of each page.
• Provide white space.
• Avoid double columns.
• Don't fold or staple your résumé.
• Use abbreviations carefully.

Remember, the purpose of the résumé, is to get you a job interview. But first you must get your résumé past the computer.

If you need more help with résumé writing and preparation, there are numerous books, pamphlets, and seminars to aid you in the process. Consult your learning resource center, student resource office, or instructor.

A résumé cannot go out alone; it must be accompanied by a **cover letter** that tells the individual receiving the résumé why it has been sent. The cover letter, written in standard business-letter format, is an opportunity to (1) direct the résumé to a specific person, (2) provide additional information about yourself relating to a specific position for which you are applying, (3) describe briefly what you know about the job and employer, and (4) indicate what follow-up action will be taken on your part.

One of Bob's cover letters (unlike the résumé, each cover letter is different) is shown in Figure 9.3. Note the following:

1. Applicant's address and date of letter are given.
2. The prospective employer's name, title, and address are provided.
3. The salutation addresses the specific contact at the company.
4. The *opening paragraph* states why you are writing, names the position or type of work for which you are applying, and mentions how you heard about the opening.
5. The *middle paragraphs* explain why you are interested in working for this company and highlight one or two points from your résumé (but do not duplicate the résumé). Include phone numbers (and e-mail address, if you have one) at some point in the letter to be sure the prospective employer knows how to reach you.
6. The *closing paragraph* has an appropriate closing to pave the way (or ask) for the interview.
7. Be sure to both sign and type your name.

Remember, your résumé and cover letter together are your written introduction to an employer. Make them both flawless.

When an employer responds to your cover letter by inviting you for an interview, the first thing you will be asked to do is complete a **job application.** Like the résumé and cover letter, it must be done right—that is, it must be filled out correctly, neatly, and completely. As with the résumé, it is not just what the application says, but how information is placed. Care in following directions is essential, as it indicates to many employers your ability to read and follow instructions as well as how careful you are.

The best way to fill out a "perfect" application is to complete a *master application* ahead of time. Find the most detailed employee application available and spend an hour or two filling it out. Then, bring it along with you to every employer, where you will use it as a guide in completing the individual application. With the use of a master application, which has been checked and rechecked for accuracy, spelling, correct addresses,

23052 Bretton Place
Los Angeles, CA 90015
April 3, 2001

Mr. John Jones
Personnel Director
ABC Electronics, Inc.
Los Angeles, CA 90012

Dear Mr. Jones:

My neighbor, Joe Heart, a supervisor at your organization, told me that there is an entry-level electronics technician position available at ABC Electronics. I would like to be considered a candidate for the position.

I will graduate in June from Los Angeles Trade Technical College with an Associate in Science degree in electronics technology. I feel my technical training, work experience, and school club activities have helped me develop the skills necessary to work with electronics equipment and relate well to customers and fellow employees. I am particularly interested in product service, an area in which your company has a fine reputation.

I have enclosed my resume for your review. Should you need more information, I will be happy to provide it. You may reach me at 213/555-6455 during the day and 213/555-8090 in the evening. My e-mail address is parksele@aol.com.

I will be in touch with you within the next two weeks to request an interview. I look forward to meeting you.

Sincerely,

Bob K. Parks

Bob K. Parks

Enclosure

Figure 9.3
A cover letter.

hiring and termination dates for previous employment, etc., you don't have to drag along a briefcase full of documents to be fumbled through. You'll have all the information you need in one convenient place.

After completing the job application, but usually prior to the job interview, you may be asked to take a company **test.** There are two types, formal and informal. If a test is taken at all, you will certainly be given the former. You may not, however, even know you're getting the latter.

The written test, most likely multiple choice, tends to be heavy on basic electronics, contains some math, and requires the knowledge of Ohm's law, Kirchhoff's laws, etc. Questions on mechanics are often included, as are items related to employee safety. Thus the test, whether 15 minutes or 2 hours long, is very similar to the type you have been taking in school. Indeed, it may have been prepared by an electronics instructor. Since you're fresh out of school, you should have the advantage in taking and doing well on such tests. If you remember to apply the test-taking strategies you have been using all along, you will do fine.

The informal test, when given, is often in the form of a tour of the facilities, with questions to follow. Do not make the mistake of thinking that such a tour is just a chance for the employer to show off the company. There is often more to it than that, with the tour guide later asking you questions on what you saw and what was said. It's a chance for the employer to see how fast you pick things up—how much of a "quick study" you are. As Keith Bilby of Anheuser-Busch Brewery in Van Nuys, California, said: "I take a prospective employee on a standard 1-hour tour of the plant, the entire beer-brewing operation. Later on, during the interview, I refer back to the tour to see how much the applicant picked up and retained. It gives me another input on his or her capabilities." The moral of the story is simple: Do a little research before you arrive at the plant to fill out the job application, take a company test, or have a job interview.

Having a well-prepared résumé is important, filling out an employment application correctly is essential, passing the company test, if given, is a must, but the **job interview,** especially today, is vital—it, more than anything else, gets you the job. Why is this so? It is during the interview that the employer sizes you up, probes in questionable areas, and gets to see what you are really like. Résumés are often inflated and applications can be padded, but the interview, as much as anything, tends to "tell it like it is." While the résumé and application will get you in the door, it is the job interview that will determine whether or not you stay there.

All this may sound a bit harsh, especially when you consider that the interview is a two-way street; you're deciding on the company as much as they are deciding on you. Keeping that factor in mind will allow you to relax somewhat, take things in stride, and be yourself.

Yes, be yourself and project as much confidence as possible. Confidence is your key to success in the interview stage. Confidence can be thought of as having two aspects: technical and personal. The former is based on your electronics know-how; it comes from taking the right classes, getting good grades, and doing the "after-school electronics" talked about in the previous chapter. Personal confidence has to do with personality and how well you communicate and relate to people. It is strengthened by practice in working with others, talking and listening to them, and going through mock interviews. Both confidence factors need to be developed and nurtured from the moment you start your electronics education, if not sooner. In a sense, the interview is the culmination of those efforts, your "moment of truth." With the right training and the right attitude, your interview need not be a hurdle to overcome, but a stepping-stone to the job you want. Let's see, then, what it takes to impress your future employer by examining the key elements in the *before, during,* and *after* stages of the job interview.

There are a number of things that must be done *before the interview takes place.* First, you should investigate the company and the industry in general. Find out all you can about the company's products and services. Talk with people who work there, if possible. Also, research the industry in a generic sense. If you have an interview at a brewery, for example, 15 to 30 minutes with an encyclopedia, poring over the beer-brewing process, will be time well spent.

Second, gather up any pertinent materials you want to bring—extra copies of your résumé, work samples, list of references, etc. Never walk into an interview with nothing in your hands.

Third, organize your thoughts. Plan answers to possible questions (review your goals, needs, skills, and so forth). Find a good list of typical job-interview questions and go over it carefully.

Fourth, dress appropriately. "Dress for success," it is a phrase that has entered the mainstream. Yet, while there are jobs where impressing by dressing is to be factored in, for most people, dressing appropriately is the key, all they really need to do.

What you wear on the job, however, may be different than what you should wear to your job interview. With the former, you dress to be included, like others doing the same work. For an interview, it is always acceptable to dress upward, a little more formally than you would once on the job. At the interview, "Play it safe," says Gordon Thomas, of Menswear Solutions. "Wear something that you don't have to worry about—

no wrinkle-prone fabrics. It's fine to be overdressed at the interview, but you're in serious trouble if you are too casual."

The bottom line—dress appropriately. Once you find out what everyone else is wearing, do pretty much the same.

Fifth, be on time, although 10 or 15 minutes early is even better. *During the interview,* remember to do the following:

1. Show interest and enthusiasm. Try to relate your interests to the organization's.
2. Make points politely. Stress your best qualities and experience; note what you could bring to the position. Focus on the employer's needs, not your own.
3. Be honest. Admit when you don't know something. Remember, you're not expected to be an expert on the company's products and services, just a quick learner.
4. Be positive about former employers. Never complain about previous employers or their practices.
5. Ask for the job at the close of the interview. Of course, don't expect an affirmative response on the spot. Ask when you can expect to hear from the company.

After the interview you must do the necessary follow-up. Begin by sending a thank-you letter. Use it to ask additional questions, supply more information, underline an important point, or just remind the person who interviewed you that you're still interested in the position. Keep the letter short, address it to your interviewer, and send it right away.

If you haven't heard from a company after a set time, phone or write to ask if the job has been filled.

If you receive a letter of rejection—and you're bound to receive some—don't be discouraged. Assume that the position wasn't right for you anyway, and keep on looking.

Negotiating Salary and Benefits: Looking for Win-Win

Negotiating a salary and benefits package can be tricky; if done wrong you may lose the job offer altogether. Recruiters and career coaches say it is best to have your negotiating strategy and goals firmly in place before you interview. Once you do, follow these rules, as presented in a *U.S. News* cover story:

- *Be timely.* Wait until you have a firm job offer in hand.
- *Be enthusiastic but firm.* After you show appropriate excitement, say that you must go home and think about it overnight.
- *Be reasonable.* Check with co-workers and industry associations for a realistic salary range. You want to be in the ballpark, but not out in left field.
- *Be consistent.* Don't keep upping the ante. That's just plain insulting.
- *Be a one- or two-rounder.* Go back and forth twice, but no more. It can't become an endless cycle.
- *Be flexible.* If a salary increase just isn't going to happen, try for bonus and incentive plans.
- *Be gracious.* After all, you are going to have to live with these folks if you get the job.

Remember by taking a good look at your salary needs, understanding the market, and approaching salary and benefits negotiations that are mutually beneficial to you and your employer, your chances for success are greatly increased.

Keeping Your First Position as an Electronics Technician: The Critical Early Weeks

It has happened; you have been offered a job as an electronics technician. Congratulations! Before you accept, however, evaluate the position carefully in terms of job responsibility, working hours, pace of work, advancement opportunities, salary range, benefits, job location, and future possibilities. (What could the job lead to?) If you accept, write a confirming letter clarifying the starting date and time. You're now set; you are on your way.

When you begin work, you'll be in the "honeymoon period"—a time, from a few weeks to a couple of months, when you can be a "freshman" again. It's a vague period in which "dumb questions" from you are not only tolerated but expected. Take advantage of this opportunity to ask questions that you will be too embarrassed to put forth later on when you're expected to know the answers. I go through a similar experience every semester with a new class. I have about two weeks to learn the names of all my students. If I am still trying to sort out Jeff from Brett or Charlene from Melissa by the third or fourth week, my students will begin to wonder how much I really care.

The first few weeks on the new job are also a time to develop a reputation, a period when your co-workers and supervisors find out who you are and what you're all about. It's a time to identify with the 20 percent of the work force that is doing 80 percent of the work. You will want to hang around with them, be seen with them, work with them, and become one of them. They are the winners—the group you want to be a part of.

Throughout this period, never lose sight of what you are there for—to earn a profit for the company. While at first your productivity will be low, eventually it must become high enough to justify your presence in the organization.

Keep in mind the "no-no's," the things most likely to upset any employer. According to a national survey, here is a list of the top ten negatives:

1. Dishonesty and lying
2. Irresponsibility, goofing off, and attending to personal business on company time
3. Arrogance, egotism, and excessive aggressiveness
4. Absenteeism and tardiness
5. Not following instructions
6. Ignoring company policies
7. A whining or complaining attitude toward the company or job
8. Absence of commitment, concern, or dedication
9. Laziness and lack of motivation or enthusiasm
10. Taking credit for work done by others

On the positive side, be realistic, be patient, be alert, show initiative, be cooperative, be conscientious, be mature, and be inquisitive. And above all, keep on learning—on the job, in school, or on your own. You are entering a field that is changing so drastically that what you know today will be obsolete tomorrow. Yet, electronics is filled with so much excitement, wonder, and promise that a long and rewarding career is ahead for you. Reach out and make the most of it. Good luck.

Summary

In Chapter 9, we examined the advantages students have in seeking hi-tech jobs. We explored effective job search strategies. And we saw how to write an effective cover letter and résumé. We then examined ways to "ace" the all important job interview and how to prosper during the first few weeks on the job. We concluded with a look at ten "no-no's" likely to upset any employer.

Review Questions

1. Being right out of school, you're ready for more school; you are "_____ ready."

2. According to data presented in the chapter, _____ percent of job-seekers apply directly to an employer.

3. Used by both job-seekers and companies looking for employees, the _____ is a must job-search "engine."

4. The whole point of any job search strategy is to _____ one.

5. Your _____ is your personal inventory and formal introduction to a potential employer.

6. A resume must not be sent out alone; it must be accompanied by a _____ _____.

7. The best way to fill out a "perfect" job application is to complete a _____ _____ ahead of time.

8. The _____ _____, more than anything else, gets you the job.

9. The first few weeks on a new job is a time to develop a _____, a period when your co-workers and supervisors find out who you are and what you're all about.

10. Above all, on a new job you must keep _____.

Individual or Group Activities

1. As a group, prepare a list of nonelectronics industries that would need the services of electronics technicians. Then make a second list of specific companies in your area that might hire such techs. Finally, create a third list of contact persons in each company.

2. In the chapter, we list and discuss a half-dozen key job-search methods. Try to expand the list to a dozen or more, adding comments as you do. Prepare your own job-search plan, similar to the one in Figure 9.1.

3. Prepare your résumé and have it critiqued by fellow students. Upon conclusion, each student should have a "perfect" résumé ready for distribution.

4. Get a job application. Select the best of all those obtained by the class to serve as the *master application*. Photocopy enough copies for each student, and complete the application.

5. Form small groups and conduct mock interviews. If possible, have them videotaped. Have one student play the role of interviewer, another, the interviewee. Rotate roles among group members. In the process, prepare a list of interview questions and responses. Have the entire group critique each session.

Issues for Class Discussion

1. Discuss how students who have just graduated can compete with more experienced technicians in the search for jobs. Are there particular industries or particular companies that are more amenable to entry-level techs? Just what is an entry-level technician?

2. Prepare a list of "interview don'ts": mannerisms, attitudes, dress, or responses to questions that are sure to offend an interviewer. Discuss each item in class.

3. Ten no-no's, or negatives, that are likely to upset employers are listed in the chapter. Discuss the ramifications of each. Expand the list if you can.

4. The new employee tends to "show off a bit," to let the employer know he or she didn't make a mistake in hiring the employee. Yet this is a time to ask plenty of questions, even "play dumb," if necessary. Discuss how these two conflicting inclinations can be resolved to the new employee's benefit.

5. Invite a guest speaker from industry to discuss interview techniques. This person should be from the human resources department or, ideally, an individual who interviews and hires technicians.

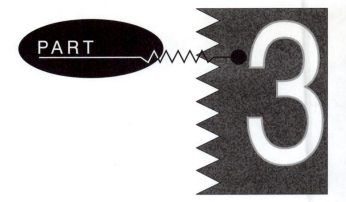

PART 3

"Making Like" an Electronics Technician

Having read Parts I and II, you should have a good idea of what it means to be and to become an electronics technician. In Part III, we'll help you act like an electronics technician, while you're still in school. We want you to get started doing electronics.

In Chapter 10, "Breadboarding: A Half-Dozen Ways to Go," we summarize the characteristics of six popular breadboarding approaches. You'll get acquainted with (1) Cut-Slash-and-Hook, (2) Solderless Circuit Board, (3) Wire Wrap, (4) Universal Printed Circuit Board, (5) Electronic Network Designer, and (6) Surface Mount Component Breadboarding. After completing Chapter 10, you'll thoroughly understand today's breadboarding methods and how to select the appropriate one for circuit experimentation.

In Chapter 11, "Creating Troubleshooting Trees: Your Key to Circuit Understanding and Repair," we look at how to create troubleshooting trees for any projects. When your circuit won't squawk, light up, or put out the required voltages, you need to come up with a detailed, project-specific troubleshooting plan, one designed to take you step-by-step through every possible contingency. In this chapter, we'll show you how to do just that.

With Chapter 12, "Transducers: The Ins and Outs of Electronic Circuits," it's on to a little electronics theory. We will examine what it takes to go from an electronic circuit to an electronic system. We'll see how input transducers convert sound, light, heat, chemical reaction, and motion into electricity to be manipulated by a circuit. We will then examine output transducers that reverse the process, converting electricity back into light, heat, and motion.

In Chapter 13, "A Short Course on Active Solid-State Components: The Big Half-Dozen," we acquaint you with today's electronic devices. You'll get an introduction to diodes, transistors, silicon controlled rectifiers (SCRs), triacs, voltage regulators, and optical couplers. Reading this chapter won't make you an expert, but it will get you heading in that direction.

Next, in Chapter 14, we examine a topic that every electronics student and technician can never know too much about. In "Electronics Safety— Where the Danger Lies," we explore, in a do's-and-don't's format, how to keep yourself safe, not only from electricity, but other dangers a home- or school-based work environment can generate.

In Chapter 15, "Getting Started With SMT: Surface Mount Technology for the Electronics Experimenter," we examine a packaging revolution that has all but replaced traditional component configuration. We see how such components can be installed and removed using ordinary hand tools, albeit, in a nontraditional manner. Upon completing Chapter 15, you'll know how to handle and work with SMCs.

In Chapter 16, we take a look at "Virtual Electronics," how project design, simulation, and analysis takes place on a computer. We examine a typical software package, Multisim V6.2, from Electronics Workbench.

Finally, in Chapter 17, "Useful Electronic Projects You Can Build," we provide the schematic drawing, parts list, project description, and construction hints for 12 electronic projects. You can get started building a Logic Probe, Carport Night-Light Controller, Pushbutton Combination Lock, or Tone-Burst Generator, to name just a few projects. It's all designed to get you doing, and thus learning, electronics.

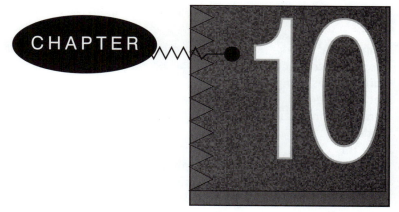

CHAPTER 10

Breadboarding:
A Half-Dozen Ways to Go

Objectives

In this chapter you will learn:

- How the cut-slash-and-hook (CSH) method of breadboarding is used with circuits requiring large currents.
- How solderless circuit boards allow for solderless assembly of modern circuits.
- How wire-wrapping results in a secure, solderless connection.
- How universal printed circuit boards are designed to make wire-wrapping more convenient.
- How the Electronic Network Designer (END) product combines the advantages of solderless circuit board and universal printed circuit board.
- How, using Surfboards, it is possible to breadboard flea-size Surface Mount Devices (SMDs).

If you are to learn electronics, you must *do* electronics. That means grabbing a handful of electronic components, a schematic drawing, a few hand tools, and a **breadboard.** Then you sit down and assemble electronic circuits (projects). True, you'll make mistakes, and much of what you build simply won't work—at least not at first. But as you progress, connecting wires, soldering components in place, reading the schematic, you will learn. And as you gain knowledge and experience, your confidence to do electronics grows. You'll begin to appreciate what it means to be an electronics technician: one who installs, maintains, and repairs electronic products.

In this chapter, we will summarize the characteristics of six popular breadboarding approaches. Starting with (1) **Cut-Slash-and-Hook,** a tried and true technique for connecting heavy duty components, we'll then proceed to (2) **Solderless Circuit Board,** (3) **Wire Wrap,** (4) **Universal Printed Circuit Board,** (5) **Electronic Network Designer (END),** and (6) **Surface Mount Component Breadboarding.** After completing Chapter 10, you'll understand today's breadboarding and how to select which method, or combination of methods, is appropriate for turning a particular design into a working circuit. (For a discussion of virtual electronics, or breadboarding on the computer, see Chapter 16.)

Breadboarding, a circuit assembly system that allows components and interconnections to be assembled and changed in their design stage easily, has been going on at least since the early Atwater Kent radios were assembled on actual breadboards, using nails and bare wire, back in the 1920s. Today, nails, and, to a large extent, wires, are out. Breadboarding itself, however, is still very much in—with a host of methods we will examine in a moment.

Before we get started, though, it's important to remember two points: First, there is no right or wrong procedure; one particular technique is not necessarily better than any other. Which one you choose depends on the type of circuit you're building. Second, you may combine different methods to meet particular needs. This is often done with the wire-wrapping and universal printed circuit board approaches, for example.

Cut-Slash-and-Hook (CSH): Tried and True

CSH has been around for decades and is used at the breadboarding, prototyping, and production stages. In fact, there was a time, not that long ago, when the *only* way to build anything in electronics required cut-slash-and-hook, or what is generally referred to as point-to-point wiring. Today, CSH is used at the breadboarding stage only with circuits requiring large amounts of current, such as big power supplies and audio amplifiers. In such cases, it's a method that still offers many advantages.

With CSH, you take a piece of wire, cut it to length, strip it on both ends, and hook it around a terminal post before soldering. The same thing is done with component leads, but of course the lead is not stripped of insulation.

Using the CSH method, **terminal strips** or **push-in terminals** are attached at various locations to a breadboard material usually made of perforated phenolic (plastic) ⅟₁₆ inch or ³⁄₃₂ inch thick (Figure 10.1). The terminals are either fastened with screw and nut or pressed into place. A wire or component lead is then simply hooked around the terminal and soldered in place.

Electronic components too heavy or bulky to tie directly to a terminal can be screwed or strapped down onto the breadboard. Wire of almost any gauge is then used to connect the component terminals to the terminal strip.

The CSH method still has advantages. It's easy to do, and the skills required to perform cut-slash-and-hook are quickly learned. However, the CSH technique has some major drawbacks, too. It's very time consuming,

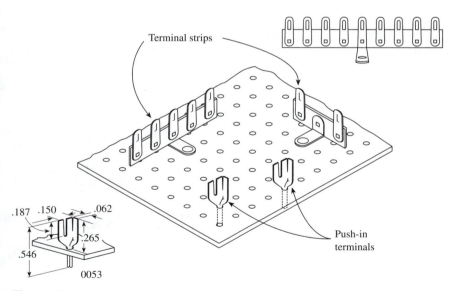

Figure 10.1
CSH breadboarding materials.
Source: From *Electronic Project Design and Fabrication,* Fifth Edition (p. 110), by R. A. Reis, 2002. Upper Saddle River, NJ: Prentice Hall. Copyright 2002 by Prentice Hall Publishing Company. Reprinted by permission.

it takes up a great deal of space, and with CSH it's difficult to change things. But perhaps the biggest disadvantage has to do with compatibility. The CSH procedure of breadboarding is totally unsuitable for the assembly of modern electronic circuits, especially those using integrated circuits. Digital, and many analog, circuits simply cannot be breadboarded with this method.

Solderless Circuit Boards: Breadboarding Modern Circuits

A method actually designed with today's through-hole IC technology in mind has been around for some time. Known by the generic term *solderless circuit board*, it's made up of a plastic breadboard with hundreds of holes (Figure 10.2). The boards come in many different sizes, ¼ × 2 × 5-inch being the most popular. They can be snapped or butted together to increase the working surface. The holes are placed in a rectangular grid at regular intervals, 0.1 inch apart. Some boards have number and letter designations for each column and row of holes to aid in component lead and wire placement.

Each hole has a tiny metal lug inside. The lugs are connected underneath in small and large groups. The usual pattern is to connect 5 lugs in

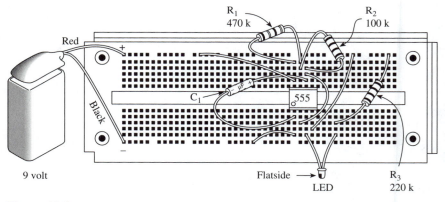

Figure 10.2
Assembling on solderless circuit board.
Source: From *Electronic Project Design and Fabrication,* Fifth Edition (p. 113), by R. A. Reis, 2002. Upper Saddle River, NJ: Prentice Hall. Copyright 2002 by Prentice Hall Publishing Company. Reprinted by permission.

a vertical column of small groups and 25 to 40 or more lugs in large horizontal groups known as *buses*.

The system works like this: Two or more component leads or wires inserted in the same group are "shorted" together by the lug-connecting strip underneath. The wire (or component lead) can usually be pushed in or pulled out with your fingers, although a pair of needle nose pliers is handy.

One important feature of every solderless circuit board, regardless of the manufacturer, is the center channel. Integrated circuits are straddled across the channel so one row of IC pins does not "short" to the row on the opposite side.

In addition to the solderless circuit board itself, the only other material used with the system is hookup wire. You can buy a kit of precut, pre-stripped single-strand 22 gauge wire from a number of distributors. Or you can use commonly available 22 to 30 gauge telephone wire and do the cutting and stripping yourself.

The advantages of breadboarding with solderless circuit board are numerous: it's a great time-saver, circuit changes are quick and easy, you can use components over and over again, and everything is laid out in front of you.

In all fairness, there are a few disadvantages, too. Some technicians who do a great deal of radio frequency work point out their circuits need large amounts of shielding, which is difficult to create with this type of breadboard. And, of course, the circuit assembly itself can be somewhat flimsy. You do have to be careful how you handle the finished breadboard. Yet when you compare the important advantages against these limited drawbacks, you quickly see solderless circuit board makes for an ideal breadboarding approach to be used with modern electronic circuits.

Wire Wrap: Twisting the Night Away

Wire wrapping is a breadboarding method that not only is compatible with modern electronic circuits, but also provides a strong and secure bond between components—without the use of solder. Although circuit construction is not as fast as with solderless circuit board, wire wrapping takes considerably less time than would be required to design and produce a dedicated printed circuit board.

In the wire-wrap technique, connections are made by simply wrapping wire ends around rectangular or square terminals. The key is to use terminals with right-angle corners. Using a special (though inexpensive) wrapping tool (Figure 10.3a), you tightly wrap a piece of 28 or 30 gauge wire around a terminal six to eight times (Figure 10.3b). The wire "bites" into the corners of the terminal. As a result, the two materials are actually "welded" together to form a solid connection. The connection is even better than solder because no cold joints are possible. Unwrapping is as easy as wrapping. Just twist the tool in the direction opposite to that used when wrapping the wire.

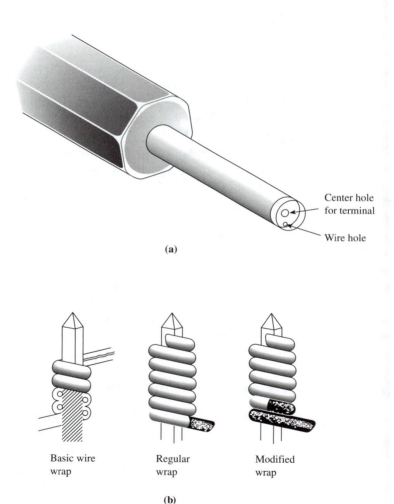

Center hole
for terminal

Wire hole

(a)

Basic wire
wrap

Regular
wrap

Modified
wrap

(b)

Figure 10.3
Wire-wrap techniques.
Source: From *Electronic Project Design and Fabrication,* Fifth Edition (pp. 114, 117), by R. A. Reis, 2002. Upper Saddle River, NJ: Prentice Hall. Copyright 2002 by Prentice Hall Publishing Company. Reprinted by permission.

In addition to the wire wrap tool, to do wire wrapping you are going to need perforated phenolic board with holes spaced 0.1 inch apart. Into these holes will go wire-wrap IC sockets and various wire-wrap posts. The sockets have long rectangular pins to take two or three levels of wrapped wire. The posts are square throughout their length and have a trifurcated clip on the top. Electronic components with round leads are soldered to the clip, and the posts are joined by wire wrapping. Remember, you cannot wire-wrap round component leads.

Another approach to handling small components, such as transistors, low-wattage resistors, and disc capacitors, is to solder them directly to a **parts carrier,** or **header.** The header is pressed into a wire-wrap IC socket, which is then wrapped in the usual way to make the circuit interconnections.

When wire wrapping, you can't use just any wire; it has to be 28 or 30 gauge, bare or insulated, and if insulated, stripped back on both ends. The wire comes in both rolls and precut/prestripped lengths. Many colors are available, the most popular being red, blue, white, black, and yellow.

Wire wrapping has the advantage that, like the solderless circuit board method, it's compatible with modern circuits. But unlike the latter, this circuit construction method results in solid, secure connections.

Like all breadboarding systems, wire wrapping does have one or two disadvantages. Wire wrapping materials tend to be expensive. Furthermore, the process is time-consuming, especially when compared with breadboarding on solderless circuit board. But if you are looking for a stable, more or less solderless circuit construction method compatible with today's components, give wire wrapping a try.

Universal Printed Circuit Boards

Universal printed circuit boards are not what you might think. They are actually more like a wire-wrap board than a printed circuit board. They are designed to make wire wrapping more convenient. Why they are called universal printed circuit boards nobody seems to know.

Universal printed circuit board is basically wire-wrap board in which rings of tinned copper surround holes on one side of the board (Figure 10.4). On some boards, strips of copper tie two or more holes together. Also included are bus strips, which can run the entire length of the board, at the edges, or anywhere in the center. Even boards with odd-shaped patterns are available, where holes are connected with copper strips in various square, triangular, or rectangular configurations.

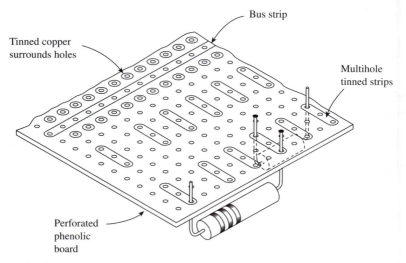

Figure 10.4
Installing components on a universal circuit board.
Source: From *Electronic Project Design and Fabrication,* Fifth Edition (p. 118), by R. A. Reis, 2002. Upper Saddle River, NJ: Prentice Hall. Copyright 2002 by Prentice Hall Publishing Company. Reprinted by permission.

The purpose of adding all this copper to an otherwise "clean" perforated phenolic board is twofold. First, the individual copper pads allow for soldering wire-wrap sockets and terminals in place. Just a dab of solder and the pins and posts are secure. Second, the multipad connections can be used to tie components together. In some cases you don't even need wire-wrap terminals. Just insert the component leads from the plain side of the board and solder them to common pads (Figure 10.4). Bus strips, of course, allow you to connect many wires and component leads to a common circuit point.

Electronic Network Designer (END)

Wouldn't it be great if the advantages of solderless circuit board and universal printed circuit board could be combined in a single product? Well, such a device does exist, having been developed by Platform Systems, Inc., of Calgary, Canada (403/245-2755). Known as the Electronic Network Designer (END), it fuses the quick assembly and easy exchange of components exemplified by solderless circuit board with the ruggedness and dependability of a universal printed circuit board. Using the END

Figure 10.5
Electronic Network Designer.
Source: Platform Systems, Inc.

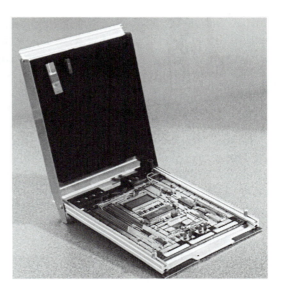

product not only saves breadboarding time, it eliminates wiring errors that inevitably crop up when a circuit first assembled on solderless circuit board is reconfigured on a universal printed circuit board.

The END consists of an aluminum frame which places a thin circuit board over a solderless circuit board-like matrix (Figure 10.5). Components are inserted through the circuit board into the sockets of the solderless circuit board–matrix. The circuit board is thin enough for short-leaded components, such as integrated circuits and SIP resistor packages, to fit snugly into the matrix pins. The circuit is prototyped as usual.

When the design is complete, the designer can tidy up the wiring and make minor changes. Next, the cover is closed and the case is flipped upside down. The bottom panel is lifted, freeing the components from the socket pins and exposing the component leads for soldering. Diced conductive foam holds the components in place for soldering. When complete, the tested and soldered board is removed from the frame.

As you can see, the END system takes breadboarding one step further, to true circuit prototyping.

Surface Mount Component Breadboarding

Of the five breadboarding approaches discussed so far, four are out of the question when it comes to working with tiny, rice-grain size surface mount components. There is simply no way you're going to breadboard a surface mount technology (SMT) circuit using the cut-slash-and-hook, wire wrap, universal printed circuit board, and Electronic Network Designer methods. The only approach even remotely possible is solderless circuit board. But, remember, with SMT, you're working with components that for the most part have no leads. Or if they do, the leads are not spaced or shaped "right." Clearly the solderless circuit board technique, by itself, isn't going to work. Someone has to come up with a better idea.

Fortunately, someone has! He's Rob Laschinski, president of Capital Advanced Technologies, Inc., in Carol Stream, Illinois (708/690-1696). His company has developed a unique breadboarding method specifically designed for surface mount components. It's called *Surfboards*, and you don't have to be a Beach Boy to appreciate how easy and effective this approach is.

The Surfboard concept involves use of specially designed universal PC boards with single in-line pins (SIPs) spaced 0.1 inch apart (Figure 10.6). The pins fit into a solderless circuit board. Thus, in a sense, this method, too, combines the universal PC board and solderless circuit board concepts.

The boards themselves incorporate foil patterns that accommodate a wide variety of component mounting footprints, both discrete and IC. Several device sizes can be mounted on the universal foil pattern (Figure 10.7).

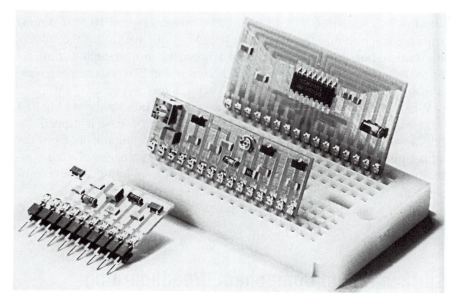

Figure 10.6
Surfboards.
Source: Capital Advanced Technologies, Inc.

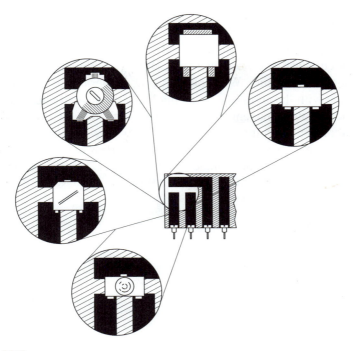

Figure 10.7
Placing SMCs on Surfboards.
Source: From *Electronic Project Design and Fabrication,* Fifth Edition (p. 291), by R. A. Reis, 2002. Upper Saddle River, NJ: Prentice Hall. Copyright 2002 by Prentice Hall Publishing Company. Reprinted by permission.

There are a number of ways to solder the surface mount components to the boards. You can hold the components in place temporarily with ordinary clear tape, then solder the component terminals to the foil pads, after which you then remove the tape. Or, you can glue the components down, then solder as above. Finally, you can use the tag solder method. Here you would first flow-solder on one pad. Then, using tweezers, hold the SMC in place and reflow the solder. You then solder the other pad in the normal manner. ICs are soldered in place using any of the three methods.

For a more detailed discussion of SMT construction methods, with an emphasis on SMD attachment and removal, see Chapter 15.

There you have it—six ways to breadboard, six ways to prove out a design before committing to a full-blown prototype or mass production. As a final note, keep in mind you can mix and match these methods as necessary. You don't have to commit to one over the other. Use your creativity to produce the most effective working circuit, as you *do* electronics.

Summary

In Chapter 10, we looked at six popular breadboarding approaches. We saw how cut-slash-and-hook is a tried and true technique for connecting heavy duty components. We looked at how solderless circuit boards are compatible with through-hole IC technology. We investigated wire-wrap, a solderless, but permanent, breadboarding method. And we saw how the universal printed circuit board is basically a wire-wrap board with copper traces. The END product, that combines solderless circuit board and the universal circuit board advantages, was explored. Finally, we examined Surfboards, a great way to prototype SMDs.

Review Questions

1. With the cut-slash-and-hook method of breadboarding, you
   _____ a piece of wire to length, _____ it
   on both ends, and hook it around a _____ post before
   soldering.

2. With solderless circuit boards, holes are placed _____
   inch apart.

3. In the wire-wrap breadboarding method, connections are made by
   simply wrapping wire ends around _____ or
   _____ terminals.

4. With wire-wrapping, the wire and the terminal are
   "_____" together to form a solid connection.

5. The universal printed circuit board is more like a
   _____-_____ board than a printed circuit
   board.

6. The END product not only saves time, it eliminates _____
   errors.

7. The END product consists of an aluminum _____
   which places a thin circuit board over a _____ circuit-
   board matrix.

8. The Surfboard concept involves use of specially designed universal
   PC boards with _____ in-_____ _____
   (SIPs) placed 0.1 inch apart.

9. Using the Surfboard method, you can hold components in place
   temporarily with ordinary clear _____.

10. In breadboarding, you don't have to commit to one
    _____; you can mix and match various approaches.

Creating Troubleshooting Trees:
Your Key to Circuit Understanding and Repair

Objectives

In this chapter you will learn:

- About the primary cause of project failure, incorrect component placement.
- How a troubleshooting tree works like a traditional computer-type flowchart.
- About the five elements of a troubleshooting tree.
- How a 5- and 9-volt power supply works.
- How the power-supply troubleshooting tree is created.
- How to create your own troubleshooting tree.
- How a 3-channel color organ works.

We hope that, after reading Chapter 10, you're ready to try your hand at project breadboarding. That's great! But be prepared for trouble. No matter how careful, how methodical you are in project construction, sooner or later something is going to go wrong. When it does, you'll need to troubleshoot the project. Don't dismay. Learning how to troubleshoot and repair a nonworking electronic device is at the heart of what electronics technicians do. When your project doesn't work—be glad. Now, in addition to learning how to build the project, you will learn how to fix it.

When your project won't squawk, light up, or put out the required voltages, you need to come up with a detailed, project-specific troubleshooting plan, one designed to take you step-by-step through every possible contingency. You need a *project troubleshooting tree*.

How do you get such a tree? In most cases, you'll have to "grow" it yourself. Doing so has its advantages. By designing your own troubleshooting tree, you will get to know a project's theory of operation down to the last detail. Furthermore, in the process you'll be creating a blueprint to hone your troubleshooting skills for all the electronic circuits to come.

In this chapter we'll show you what a project troubleshooting tree is, explain why it's so useful, and discuss its basic elements. We'll then create a troubleshooting tree for a Power Supply Project. At this point you should be ready to give it a try yourself. But, first, we'll give you some "cultivating" tips. Then we'll supply the schematic and discuss the theory of operation for a 3-Channel Color Organ Project. Your challenge will be to create its troubleshooting tree.

Before You Plant the Seed

Before you plant the seed and start cultivating a troubleshooting tree for a just-completed project, you'll want to spend a moment in preliminary project inspection and check-out. A moment spent now will save much time in detailed troubleshooting later on.

During inspection, look for three "corrects:" correct component placement, correct wiring, and correct soldering.

The primary cause of project failure is incorrect component placement. Either the wrong component is inserted in a given location or a polarized component is placed in the right location but inserted in the wrong direction. Either condition could result in project malfunction. Examine the circuit scrupulously; make sure all components are positioned correctly.

The second most prevalent cause of project failure is incorrect wiring: missing connections, wrong connections, and broken connections. The latter is a particularly annoying problem, showing up frequently in the form of hairline cracks in the copper traces of printed circuit boards. If you suspect a given trace, check it with a continuity tester or an ohmmeter.

A third, very critical check requires you to inspect for poor solder connections. Cold-solder joints will prevent adequate contact between component leads or wires and PC board traces. Be sure all solder connections are shiny rather than dull gray. Resolder any that aren't.

After you have completed these preliminary inspections, turn on your project. If you suspect a malfunction, try using four of your five senses to check for the more obvious problems.

By using the senses of sight, hearing, smell, and touch, you can do a surprisingly thorough project check-out. With sight, you can see if a component has fractured or burned out. By listening, you can detect unusual sounds, such as crackling, high-pitched whines, and sizzling. With smell, you can sniff for leaky transformers or capacitors, burning coils, and searing plastic from "cooking" transistors and ICs. And with the sense of touch, you can feel for components that are too hot or too cold.

At this point, after you have completed your preliminary inspection and check-out, it's time to consult your project-specific troubleshooting tree.

Troubleshooting Tree Structure

A **troubleshooting tree** (also known as a troubleshooting flowchart) is essentially a traditional computer-type flowchart adapted to take a project builder step-by-step through every possible circuit problem. (You will find an example of a troubleshooting tree later in this chapter, in Figure 11.3.) When you've worked your way through the various procedures indicated, you should wind up with a fully functioning project.

A troubleshooting tree is an invaluable troubleshooting tool because it represents a logical approach to finding the cause of circuit malfunctions. It's a process that says: "Do this, and if it does not correct the problem, do the next thing on the list. Keep proceeding until the problem is solved." Troubleshooting trees always have two properties. One, they present all possible problems. Two, they present such problems, or faults, in a logical sequence. If your troubleshooting tree contains these two attributes, it's most likely drawn correctly and will function well as a troubleshooting aid.

The troubleshooting tree consists of the following five elements (Figure 11.1):

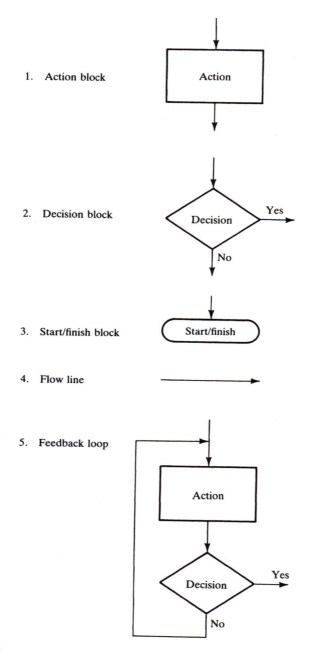

Figure 11.1

Troubleshooting tree elements.

Source: From *Digital Electronics Through Project Analysis* (p. 18), by R. A. Reis, 1991. Upper Saddle River, NJ: Prentice Hall. Copyright 1991 by Prentice Hall Publishing Company. Reprinted by permission.

1. *Action Block.* Rectangular block in which is stated an action or actions to be taken. It has one input and one output.

2. *Decision Block.* Diamond block in which a yes no question is asked. It has one input and two outputs (yes or no).

3. *Start/Finish Block.* Elliptical block in which is placed the word "start" or "finish," as appropriate.

4. *Flow Line.* Line with an arrow indicating the direction to be taken between action and decision blocks.

5. *Feedback Loops.* Formed with a flow line from a decision block output (yes or no) to an action block input.

Let's apply these elements to the construction of a troubleshooting tree for a Power Supply Project.

The Troubleshooting Tree Illustrated

As the complete schematic of Figure 11.2 shows, the power supply is designed to provide a fully regulated +5 V and an unregulated +9 V dc. Line cord PL_1 couples the 120 V ac from the wall socket, via fuse F_1 and switch

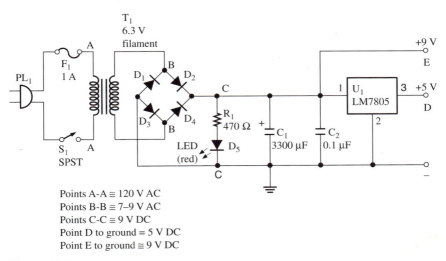

Points A-A $\cong$ 120 V AC
Points B-B $\cong$ 7–9 V AC
Points C-C $\cong$ 9 V DC
Point D to ground = 5 V DC
Point E to ground $\cong$ 9 V DC

Figure 11.2
Five-and nine-volt power supply.
Source: From *Digital Electronics Through Project Analysis* (p. 19), by R. A. Reis, 1991. Upper Saddle River, NJ: Prentice Hall. Copyright 1991 by Prentice Hall Publishing Company. Reprinted by permission.

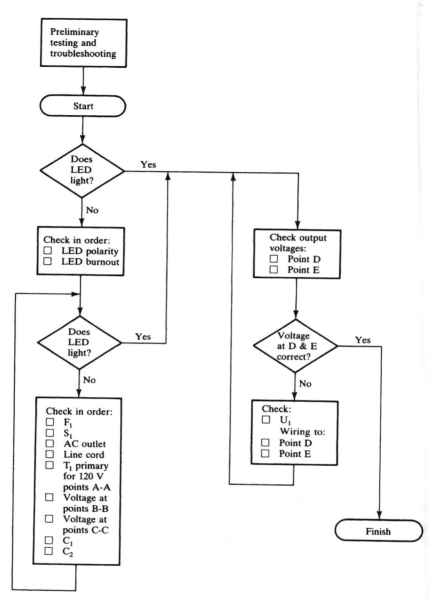

Figure 11.3

Troubleshooting tree for 5- and 9-volt power supply.

Source: From *Digital Electronics Through Project Analysis* (p. 19), by R. A. Reis, 1991. Upper Saddle River, NJ: Prentice Hall. Copyright 1991 by Prentice Hall Publishing Company. Reprinted by permission.

S_1, to the primary of transformer T_1. The transformer steps down the incoming voltage to approximately 9 V ac. Diodes D_1–D_4 produce full-wave rectification. LED D_5, with its associated current-limiting resistor R_1, gives a visual indication when the circuit is turned on. Capacitors C_1 and C_2 provide filtering and transient suppression, respectively. An LM7805 5 V regulator, U_1, ensures a steady +5 V at output D regardless of load conditions.

The troubleshooting tree for the Power Supply Project is shown in Figure 11.3. Let's follow it from start to finish.

After preliminary inspection and check-out, we start and proceed to the first decision block of the troubleshooting tree. Here we are asked: Does the LED light? If the answer is yes, we proceed to another action block, as indicated. There we are told to check for correct output voltages at points D and E. We then advance to a second decision block where we are asked: Are the output voltages at D and E correct? If the answer is yes, we head straight to the finish block—the project works. If the voltages are not correct, it's on to a new action block where we are instructed to check voltage regulator U_1 and the wiring to points D and E. After *each* check we loop back to the earlier action block, as indicated, and check for correct output voltages again. Eventually we'll get the proper voltages, break out of the loop, and proceed to the finish block.

If when we're first asked if the LED is lit, the answer is no, we proceed to a different action block, as indicated. Here we check for correct LED polarity and to see if the LED is burned out. Then it's forward to a new decision block that asks us again if the LED is lit. If it is, we proceed as delineated in the previous paragraph. If the LED is still not lighting, we advance to a final action block where we go through a series of component and voltage checks, in the order listed. After *each* check we loop back to the question: Does the LED light? Eventually the answer will be yes, and we proceed to the section on the right, where we go through the previously discussed checks.

Cultivating Tips

Here are a half-dozen "cultivating," or design, tips to consider when developing your own project-specific troubleshooting tree:

1. Always start by determining exactly what your project is supposed to do. Keep in mind, it can often be more than one thing.

For example, our Power Supply Project is actually supposed to do two things. One, it should turn on its LED indicator, and, two, it should deliver +5 V and +9 V dc at points D and E, respectively.

2. You should have at least as many decision blocks as you have things for the project to do. Looking at it another way, you will have a question for each function your project carries out. Again, examining the Power Supply Project's troubleshooting tree, we see two distinct decision blocks. (The diagram contains three, but one of the three is actually repeated.) Thus two things to do, two questions to ask.

3. Begin actual drawing with a branch or branches that lead as directly as possible to the finish, and thus a working project.

4. Place a decision block immediately following the "start" elliptical block. Be optimistic, jump right off asking if the project works.

5. One of the outputs of a decision block, usually the "no" output, often goes directly to an action block where various procedures are taken to correct the problem.

6. Keep in mind, creating a troubleshooting tree is often a matter of style. No two troubleshooting trees for the same project are likely to grow and branch in exactly the same way. The tree, like a computer program from which it derives its concept, can be tight, with only the necessary details, or open, with every step and action delineated. Again, like a computer program, some trees are simply more elegant than others. That's OK. When you design yours, just be sure the tree presents all possible faults and they're shown in a logical sequence. Elegance will come with practice.

Your Tree-Growing Challenge

Are you ready to try your hand at growing a project-specific troubleshooting tree? If so, simply pick a project and go for it. Or, we'll choose one for you. In Figure 11.4 we present the schematic of a moderately complex project, a 3-channel color organ.

The color organ is a project that flashes three strings of Christmas lights to the sound of music produced by a stereo system or FM radio. Each string of lights responds to a range of frequencies: bass, midrange, and treble. When the lights are placed behind a piece of translucent plastic, the display can be quite dramatic.

With regard to theory of operation, resistor R_1 limits current to the indicator bulb I_1. The audio input signal, connected via jack J_1, is current-limited by resistor R_2. This resistor can be bypassed by slide switch S_2, applying the full audio input signal to transformer T_1. When the audio source provides a high input signal, S_2 must be open; when a low-level signal is present, S_2 is closed. T_1 is used as a step-up transformer to provide sufficient audio levels to trigger SCRs 1, 2, and 3. Potentiometers R_3, R_4, and R_5 are used as voltage dividers, allowing for variable adjustment of the SCR trigger levels. Resistor R_6 with capacitor C_1 form a low-pass filter for frequencies below 500Hz for SCR_1 gate drive. Components R_7, R_8, C_2, and C_3 form a band-pass filter for frequencies between 500 and 3,000Hz for SCR_2 gate drive. Finally, capacitor C_4 and resistor R_9 form a high-pass filter for frequencies above 3,000Hz for SCR_3 gate drive.

Now you're ready to "grow" the project's troubleshooting tree. A solution is shown in Figure 11.5. Don't peek just yet.

Besides, your solution probably won't match the one in the figure—it may be better. Whatever the results, you're on your way to greater circuit understanding and repair for all your projects to come.

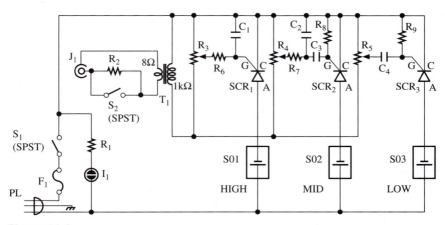

Figure 11.4
Three-channel color organ.
Source: From *Electronic Project Design and Fabrication,* Fifth Edition (p. 354), by R. A. Reis, 2002. Upper Saddle River, NJ: Prentice Hall. Copyright 2002 by Prentice Hall Publishing Company. Reprinted by permission.

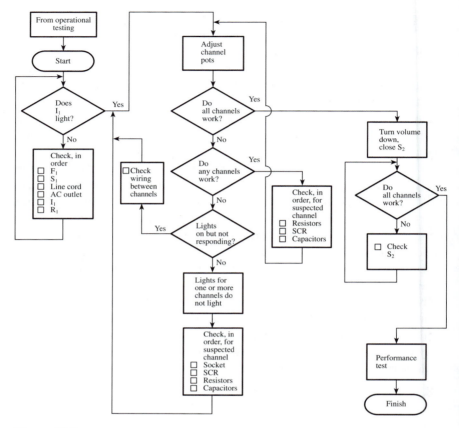

Figure 11.5
Troubleshooting tree for 3-channel color organ.

Source: From *Electronic Project Design and Fabrication,* Fifth Edition (p. 236), by R. A. Reis, 2002. Upper Saddle River, NJ: Prentice Hall. Copyright 2002 by Prentice Hall Publishing Company. Reprinted by permission.

Summary

In Chapter 11, we examined the troubleshooting tree, a device that takes a project builder step by step through every possible circuit problem. We explored the five elements that make up the tree and we looked at a half-dozen "cultivating" tips to help you "grow" the best tree possible. Finally, we probed the operation of two projects, a power supply and a color organ, and the troubleshooting trees created for them.

Review Questions

1. The primary cause of project failure is _____ component placement.

2. The second most prevalent cause of project failure is incorrect _____.

3. A third problem with project failure occurs when there are poor _____ connections.

4. A troubleshooting tree "says," "Do this, and if it doesn't correct the problem, do the _____ thing on the list."

5. A troubleshooting tree consists of five elements: an _____ block, a _____ block, a _____/_____ block, _____ lines, and _____ loops.

6. In a troubleshooting tree, a decision block is a diamond block in which a _____/_____ question is asked.

7. When developing a troubleshooting tree, always start by determining exactly what your project is supposed to _____.

8. With a troubleshooting tree, you should have at least as many _____ blocks as you have functions for the project to perform.

9. One of the outputs of a decision block, usually the "no" output, often goes directly to an _____ block.

10. Creating a troubleshooting tree is often a matter of _____.

CHAPTER 12

Transducers:
The Ins and Outs of
Electronic Circuits

Objectives

In this chapter you will learn:

- The difference between an electronic circuit and an electronic system.
- How input transducers are energy sensors.
- How photoresistors, thermistors, and piezoresistors vary their resistance with light, heat, and pressure, respectively.
- How the variable resistor/variable voltage principle works.
- How a light-controlled oscillator, automatic temperature controller, and pressure gauge illustrate the variable resistance/variable voltage principle.
- How output transducers indicate an output or take an action.
- How solenoids and motors work.

By now you should be getting "into" electronics: building and troubleshooting electronic circuits and projects. As you do, one way or another, you'll begin to pick up a little electronics theory. You will start to understand how your circuits work. Then something interesting is bound to take place. You'll come to notice a most unusual thing about electronics. You will discover that, in and of itself, electronics is practically useless.

Take the typical stereo amplifier. As a 200-watt-per-channel power amp, with its various compensation and equalizing circuits, it's an impressive unit sitting there on your shelf. It does a super job of using a weak analog input signal to manipulate a much stronger analog output signal. The amplifier is an all-electronic device—its input is electrical, its output is electrical. But that, by itself, doesn't mean much. Although it may look imposing in your wall unit, if you don't have input and output transducers (microphone, tape head, earphone, speaker) connected to the stereo, it won't sooth you to sleep or energize your next party.

We could say the same thing about a computer, if by computer we mean only the main cabinet housing the mother-board, interface cards, and power supply. This digital signal processor, with its electrical input and output signals, can't do any computing until input and output peripherals (actually transducers), such as the keyboard, disk drive, mouse, video display terminal, and printer, are connected.

Circuits, analog or digital, that manipulate current are central—they are at the core of electronics. But until we add input and output transducers, all we have is an *electronic circuit;* we do not have an *electronic system.* It's such a system, however, that makes the unit complete, with useful and exciting possibilities.

In this chapter, we'll begin by looking at the functional elements of an electronic system. Then, we will explore input transducers, in particular those that vary resistance as a function of a changing energy level. To illustrate, we'll zero in on optical, thermal, and pressure devices, examining their design and application. Next, it's on to output transducers. We'll concentrate on devices that use electromagnetism to produce motion. By the time we're through, you'll have a solid grasp of what it means to add input and output transducers to electronic circuits and, by so doing, form electronic systems.

An Electronic Circuit Does Not a System Make

Electronic circuits are a combination of active and passive components connected together on a printed circuit board. Such circuits control current by switching or regulating it. But where do these circuits get an input signal to manipulate? And what do they do with the signal after it has been

Figure 12.1
An electronic system consists
of an electronic circuit(s) with
input and output transducers.

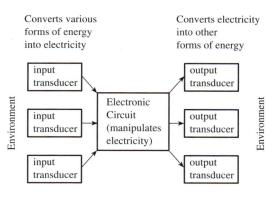

Electronic System

processed? That's where the **transducer** comes in—a device that uses one form of energy to control another form of energy.

In an electronic system, input transducers use light, heat, mechanical, and magnetic energy (to name the most common types) to affect a corresponding electrical signal. See Figure 12.1. These transducers sense the environment around them, detect the presence or changes in various "real world" energy sources, and alter an electrical signal accordingly. Examples of input transducers are photoresistors, thermistors, and piezoresistors (pressure sensors).

Electrical output transducers work in reverse (Figure 12.1). They take the switched or regulated signal from the circuit's output and use it to control other forms of energy, such as light, heat, and mechanical motion. Examples of output transducers are light-emitting diodes, piezoelectric buzzers, solenoids, and motors.

Let's begin by taking a closer look at the "ins" of electronic circuits—input transducers.

Electrical Input Transducers: The Energy Sensors

Most input transducers are basically resistors—variable resistors. But instead of using the turning of a shaft, as with a rheostat or potentiometer, to convert mechanical motion (energy) into a varying resistance, many use other forms of energy—light, heat, and pressure—to accomplish a similar conversion.

When such transducers—photoresistors, thermistors, and piezoresistors—are placed in a circuit, they act like any other resistor: they limit current and drop voltage. Since their resistance varies with a change in energy intensity, with that change comes a corresponding alteration in circuit current or voltage levels. The currents or voltages are then manipulated by the electronic circuit in an analog or digital manner. Let's look at the three types of transducers just mentioned.

Photoresistors

A photoresistor (also known as a photoconductor) varies its resistance in accordance with the amount of light striking it.

In construction, the photoresistor is formed with a thin layer of photoconductive material such as cadmium sulfide deposited on a ceramic substrate. See Figure 12.2a. Leads are attached to the photoconductive material and the entire assembly is hermetically sealed with glass. The transparency of the glass lets light reach the photoconductive material.

The schematic symbol for a photoresistor is also given in Figure 12.2a. Note the two arrows pointing inward, indicating the device receives energy, in this case, light. Sometimes the wavelength symbol is also included within the circle, as shown.

In the dark the photoresistor's resistance is quite high (few free electrons are created to cause much current flow), often several megohms. In direct, intense light, however, the transducer's resistance can drop to less than one hundred ohms (many free electrons are created to act as current carriers). Thus a photoresistor's resistance is inversely proportional to light intensity.

Thermistors

A thermistor varies its resistance in accordance with a change in temperature. Thermistors are constructed by bonding wire to various types of semiconductor materials. See Figure 12.2b. These materials produce large quantities of electron-hole pairs as temperature increases. Increased electron-hole pairs cause a drop in resistance.

Thermistors come in a variety of sizes and shapes. Some are very tiny, making them ideal for use in normally inaccessible places. Being shaped as beads, disks, washers, and rods also gives them wide application.

The schematic symbol for a thermistor is shown in Figure 12.2b. It's simply a resistor and the letter T enclosed within a circle.

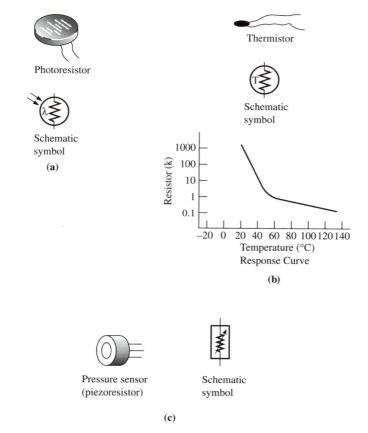

Photoresistor

Thermistor

Schematic symbol

Schematic symbol

(a)

(b)

Response Curve

Pressure sensor (piezoresistor)

Schematic symbol

(c)

Figure 12.2
Input transducers.

With a relatively small change in temperature, the resistance of a thermistor responds dramatically. The typical range is from a few ohms at 100°C to a megohm or more at 0°C. The device can provide fractional degree temperature control. Some thermistors are accurate to ±0.1°C change in temperature.

Most thermistors have a negative temperature coefficient, which means their resistance decreases as their temperature increases. We say the thermistor's resistance is inversely proportional to heat. Positive temperature coefficient thermistors do exist, but they are rare.

A major drawback with thermistors is their nonlinearity. As the response curve in Figure 12.2b shows, the lack of linearity is dramatic. In more sophisticated thermistor sensors, some form of linearization circuitry must be employed.

Piezoresistors (Force Transducers)

A piezoresistor (also known as a pressure sensor) varies its resistance in response to applied stress or strain. "Piezo" comes from the Greek, meaning "to press."

Many piezoresistive transducers are made from silicon-based semiconductor materials. If the atomic lattice structure of the material is deformed, it's more difficult for charge carriers (electron-hole pairs) to drift through. Thus the resistance of the material increases as it is deformed. We say a piezoresistive element is directly proportional to pressure. A typical pressure sensor, along with its schematic symbol, is shown in Figure 12.2c.

The Variable Resistance/Variable Voltage Principle

In Figure 12.3a are two resistors connected in series to form a voltage divider. The amount of voltage at point A is directly proportional to the ratio of R_X to R_Y, and is derived by using the formula shown in Figure 12.3b. If both resistors are of equal value, the voltage at point A is 5 volts since each resistor drops an equal amount of voltage. If R_X is twice the value of R_Y, the voltage at point A will be 6.66 volts. R_Y will drop 3.33 volts.

If either resistor in the voltage divider network is made variable, the voltage at point A will vary. If R_X is the variable resistor, as its resistance *increases,* the voltage at point A will also increase. If R_Y is the variable resistor, as its resistance *decreases,* the voltage at point A increases. Input transducers that vary their resistance on the basis of received energy can take advantage of the voltage divider principle to cause an input voltage to change. Let's see how.

With photoresistors and thermistors, resistance is inversely proportional to a rise in energy. Thus if we want to convert an increase in light or heat to an increase in voltage, we can configure a voltage divider as shown in Figure 12.3c.

With an *increase* in light or heat, the total resistance from point A to the supply voltage decreases. The *ratio* of R_1 to the combination of R_2 and

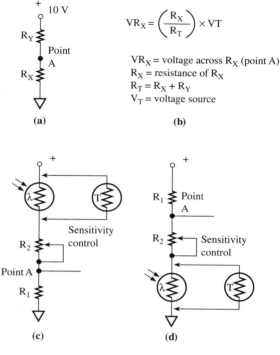

$$VR_X = \left(\frac{R_X}{R_T}\right) \times VT$$

VR_X = voltage across R_X (point A)
R_X = resistance of R_X
$R_T = R_X + R_Y$
V_T = voltage source

Figure 12.3
Voltage dividers.

the transducer's resistance increases. Thus the voltage across R_1, or at point A, increases proportionally.

If we want the reverse to happen, that is, a *decrease* in light or heat to cause an increase in voltage, we just reverse the voltage divider configuration as shown in Figure 12.3d. Now a decrease in light or heat will cause an increase in resistance between point A and ground. The ratio of the resistance of R_2 and the transducer's resistance to that of R_1 increases. Thus the voltage at point A rises accordingly.

Of course, we may not always want an energy change to cause an increase in voltage at point A. There are times when bringing the voltage at point A sharply down, close to ground, to act as a negative-edge trigger, for instance, is necessary. In that case, the circuit in Figure 12.3c becomes a "dark" or "cold" detector. When the photoresistor darkens or the thermistor cools, the voltage at point A drops. Looking at Figure 12.3d, however, we now have a circuit that detects an increase in light or heat. As the

Figure 12.4
Pressure transducers in action.

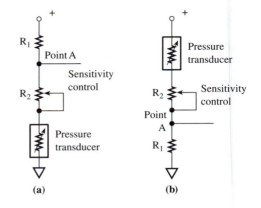

(a)   (b)

photoresistor receives more light or the thermistor heats up, the voltage at point A drops.

Since piezoresistors increase their resistance with an increase in deformity (pressure), if we want an increase in pressure to cause a rise in voltage at point A, we use the configuration in Figure 12.4a. If we want an increase in pressure to effect a decrease in voltage at point A, the circuit of Figure 12.4b is our choice.

The Principle Applied

Let's look at three circuits that illustrate the variable resistance/variable voltage principle just described.

In Figure 12.5 we see a light-controlled oscillator built around a 555 clock IC. The circuit will produce a tone from the speaker only if pin 4, the reset pin, is high. The photoresistor R_5, in series with R_6, forms a voltage divider. When light shines on the photoresistor, the voltage at pin 4 increases, and the speaker sounds. When the photoresistor is in the dark, pin 4 is close to ground, and the speaker is silent. What do you suppose would happen if the positions of R_5 and R_6 were reversed? What kind of circuit would you have?

In Figure 12.6, we have an automatic temperature controller that uses a thermistor mounted close to a heating coil. Resistors R_1 and R_2 form a voltage divider that establishes a fixed reference voltage at the "+" (noninverting) input of the comparator. Thermistor R_3 and sensitivity rheostat R_4 form their own voltage divider, the junction of which is applied to the comparator's "−" (inverting) input. As long as the voltage at the "−" input

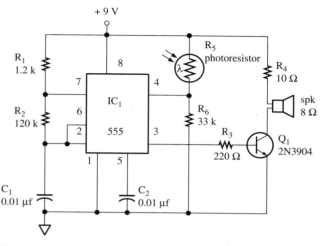

Figure 12.5
Light-controlled oscillator.

remains above the reference voltage at the "+" input, the comparator's output will be low, the triac will stay off, and so will the heater coil.

As the heater coil cools down, the thermistor's resistance increases, causing the voltage at the "–" input to fall below the reference voltage. Now the comparator's output goes high, the triac turns on, and the heater coil is energized. As the thermistor heats up, its resistance decreases, and the voltage at the "–" input rises above the reference voltage. The comparator's output again goes low, the triac turns off, and the heater coil

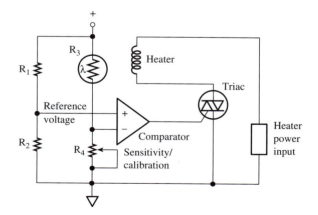

Figure 12.6
Automatic temperature controller.

Figure 12.7
Pressure gauge.

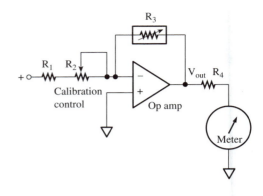

cools down. Thus we have a feedback loop monitoring the temperature of the heater coil.

In Figure 12.7, we have a pressure gauge circuit consisting of an operational amplifier (op amp) with a piezoresistive element (R_3) as a feedback resistor. As pressure increases on R_3, its resistance goes up, as does the voltage across it. Thus V_{out} increases, producing more current through the moving coil meter. Hence, the meter indications are directly proportional to pressure.

Electrical Output Transducers: Putting Electricity to Work

Electrical output transducers use the circuit's processed electronic signal to control other forms of energy, such as light, heat, and electromagnetism, the latter being used to create motion. These output transducers give electronic circuits their reason to exist.

Essentially, output transducers fall into two categories: (1) those that provide an indication of the input energy sensed, or (2) those that take action as a result of what is sensed. Examples of the former are analog meters, digital displays, LED indicators, bar graphs, plotters, etc. The latter, motion transducers, are often solenoids and motors used in countless applications, particularly those involving robots and process control equipment.

In converting electricity into mechanical motion, an intermediary energy transfer often takes place. We must first convert electricity into **electromagnetism.** Electromagnetism in turn produces linear or rotary motion. Let's take a closer look.

Electromagnetism: The Intermediate Link

Electromagnetism is magnetism produced by electricity. It is ubiquitous; it exists everywhere current does.

The electromagnetism created by every current-carrying conductor is proportional to the amount of current. More current, a stronger magnetic field; less current, a weaker magnetic field. If the current changes, so does the magnetic field. When the current is increasing, the magnetic field is expanding. When the current is decreasing, the magnetic field is contracting. Thus with an alternating current (ac) which is always changing, the associated magnetic field is always expanding or contracting, moving out from the center of the conductor or collapsing back into it.

In addition to increasing with a strong current, the strength of the magnetic field is enhanced by winding a conductor into a coil. Doing so concentrates the field, making it stronger. Generally, the more turns and the tighter they are wound, the more concentrated and stronger the electromagnetic field.

For decades engineers have been using the principle of electromagnetism to design transducers that change electricity into motion. Two such devices, solenoids and motors, deserve special attention.

Transducers That Create Motion: Solenoids and Motors

A solenoid is a simple electromagnetic transducer that translates ON/OFF electrical signals to ON/OFF (linear) mechanical movements. As shown in Figure 12.8a, in its basic form the solenoid consists of a coil of wire surrounding a moveable iron bar. The bar is located in such a way that when the coil is energized, and a magnetic field is created around it, the bar is drawn into the coil. The bar acts against the tension of a spring. Thus when current is removed from the coil, and the magnetic field collapses, the iron bar is returned to its rest position.

In industrial settings, solenoids are used for a large variety of process-control ON/OFF applications. These applications include control of hydraulic and pneumatic flow and the opening and closing of various types of actuators. Solenoids are hardworking, rugged output transducers.

Changing electrical energy into rotary mechanical motion is the job of a motor—one of the most significant inventions of the industrial age.

Motor operation depends on the interaction of two magnetic fields, as shown in Figure 12.8b. In concept, one magnetic field is developed from

Figure 12.8
Solenoids and motors.

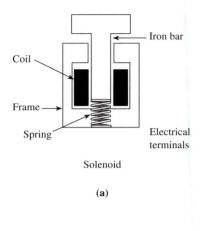

Iron bar

Coil

Frame

Spring

Electrical terminals

Solenoid

(a)

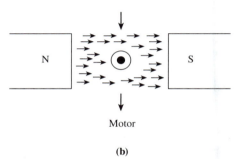

N

S

Motor

(b)

a permanent magnet, the other from an electromagnet. When the two magnetic fields interact there is a pushing and pulling effect. The electromagnet is mounted such that it is free to rotate. Thus the electromagnet moves. Electricity, through magnetism, has resulted in motion.

The uses of motors, from tiny, precision stepper motors used in disk drives, to high-speed drills employed gingerly by your dentist, are too numerous to list. Suffice it to say, without this omnipresent electrical output transducer, civilization as we know it would not exist.

Actually, we can say the same thing about electronics, if we think of electronics as a whole. Where would we be without radio, television, radar, computers, medical diagnostic equipment, avionics, automotive electronics, industrial automation? As long as we remember all of these wonders require input and output transducers to make them complete, the answer is obvious. Clearly, electronics, when we speak of electronic systems, is anything but useless.

Summary

In Chapter 12, we began by seeing how an electronic circuit does not a system make. We explored the role of input transducers: photoresistors, thermistors, and piezoresistors. We studied the variable resistance/variable voltage principle. We saw how that principle is applied to a light-controlled oscillator, automatic temperature controller, and pressure gauge. Finally, we examined the workings of output transducers, looking at those that create motion, in particular, solenoids and motors.

Review Questions

1. A _____ is a device that uses one form of energy to control another form of energy.

2. Most input transducers are basically _____ resistors.

3. A photoresistor varies its resistance in accordance with the amount of _____ striking it.

4. A _____ varies its resistance in accordance with a change in temperature.

5. "Piezo" comes from the Greek, meaning " _____."

6. Two resistors connected in _____ form a voltage divider.

7. In Figure 12.5, the _____ at pin 4 of IC_1 will vary as light hitting the photoresistor, R_5, varies.

8. In Figure 12.6, the reference voltage is _____ by the values of R_1 and R_2.

9. _____ is magnetism produced by electricity.

10. Motor operation depends on the _____ of two magnetic fields.

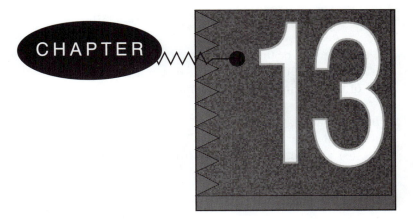

CHAPTER 13

A Short Course on Active Solid-State Components:
The Big Half-Dozen

Objectives

In this chapter you will learn:

- Why a diode is considered a one-way valve.
- How a transistor switches and amplifies current.
- How an SCR acts as a solid-state latching switch.
- How a triac operates as a back-to-back SCR.
- How a voltage regulator holds output voltage steady even with load variation.
- How an optical coupler isolates one circuit from another.

Today, solid-state (non-vacuum tube) control components are everywhere. There are rectifiers, signal diodes, zener diodes, tunnel diodes, and photodiodes. We have bipolar transistors, field-effect transistors, unijunction transistors, and MOSFETs. And, of course, there are diacs, triacs, and SCRs. The list goes on and on. It seems hard enough just to memorize the many component schematic symbols, let alone understand each device's function and use.

255

Yet it need not be that complicated or overwhelming, even for a novice. In truth there are only about a half-dozen active solid-state components. Most of the remaining "minor" components fall into subcategories of the six "majors."

In this chapter, we offer a short course on active (solid-state) components. We'll cover the basics of (1) diodes, (2) transistors, (3) SCRs, (4) triacs, (5) voltage regulators, and (6) optical couplers. This summary of component functions is designed to help you better understand and build electronic circuits and projects.

Electronic components can be divided into two broad categories: active and passive. The former are those which have gain or direct current. In this category we include diodes, transistors, SCRs, and triacs. Such components change the fundamental character of an electronic signal by switching and amplifying. Passive components, on the other hand, have no gain characteristics. Resistors, inductors, and capacitors are examples of passive devices. In this chapter we'll discuss only active solid-state components.

Diodes

A diode is a two-element (cathode and anode) component that allows current to flow through it in one direction only. The schematic symbol, pictorial presentation, and junction drawing for a typical silicon diode are shown in Figure 13.1a. The diode will conduct current when forward-biased (its cathode negative, its anode positive). It will block current flow when reverse-biased (its cathode positive, its anode negative). Thus the diode acts as a one-way valve. It's used primarily for switching, though it's also employed as a rectifier to change ac to dc.

Diodes are rated with regard to current- and voltage-handling ability. For example, a 1N4001 diode is rated at 1 A/50 V. That is, it will carry *up to* 1 ampere while handling *up to* 50 volts. Could such a diode work in a circuit requiring 500 milliamperes (a milliampere is a thousandth of an ampere) at 25 volts? Yes, of course.

What can be done with a single diode? For one thing, we can switch it in and out of an ac circuit, thus creating a dirt-cheap light dimmer, motor speed controller, or soldering iron temperature regulator (Figure 13.1b).

Perhaps the diode's biggest use is as a rectifier in power supplies to change ac to pulsating dc. Usually a bridge, consisting of four diodes, is required (Figure 13.1c). When the incoming ac waveform is positive, diodes D_1 and D_2 conduct. When the ac waveform is negative, diodes D_3

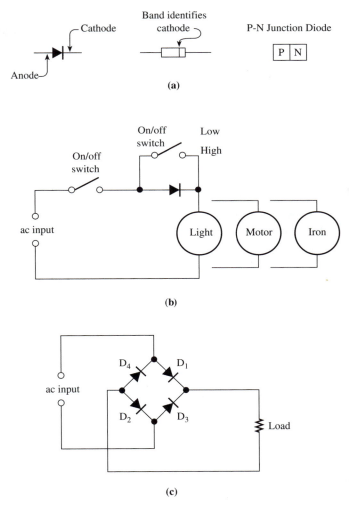

Figure 13.1
Diodes.

and D_4 conduct. During each half cycle, current flows through the load in the same direction. Thus we have dc, direct current.

Transistors

A bipolar transistor is a semiconductor component that can both switch and amplify. It consists of three terminals: emitter (E), base (B), and collector (C). There are two configurations, NPN and PNP. There are also two broad types, classified according to current-handling ability: general-purpose and power transistors. The schematic symbols for NPN and PNP transistors are shown in Figure 13.2a; the corresponding junction diagrams are shown in Figure 13.2b; and typical packaging styles are depicted in Figure 13.2c.

Current flow in a transistor is from emitter to collector (electron current flow). The amount of current that flows is determined by the polarity and extent of charge placed on the base. An NPN transistor requires a positive charge relative to the emitter; the PNP transistor requires a negative charge relative to the emitter.

A transistor is operated as an amplifier or as a switch. As an amplifier, a small varying signal placed on the base will cause a much larger varying signal to flow from collector to emitter.

When a transistor is operated as a switch, it is either at cutoff (no collector-emitter current flows) or at saturation (maximum collector-current flows).

Operating as a simple, reliable switch, the discrete transistor finds numerous applications. In Figure 13.2d we see a water-level detector circuit built around a PNP transistor. A negative base bias is set with a voltage divider consisting of resistors R_1 and R_2. Current stabilization is maintained by R_3. When water shorts out the water probes, the bias circuit is complete, collector current flows, and the relay is activated. Removing the water probes from the water causes the relay to de-energize.

A transistor amplifies ac through the use of its input circuit and output, or load, circuit. In Figure 13.2e we find a basic common-emitter circuit, centered on a general purpose NPN transistor. The value of base resistor R_b and the collector voltage V_{cc} determines the base current and bias voltage. The incoming signal adds or subtracts from the emitter-base bias voltage.

When the input signal increases in a *positive* direction, it adds to the emitter-base forward bias, causing the base and collector currents to increase. Thus the emitter-collector resistance R_{ce} decreases. With less resistance, less voltage will be dropped across the output circuit resistance. As a result, the output signal will *decrease*.

When the input signal increases in a *negative* direction, it subtracts from the emitter-base forward bias, causing the base and collector currents to decrease. Now the emitter-collector resistance R_{ce} increases. With more resistance, more voltage will be dropped across the output circuit resistance. Hence, the output signal will *increase*.

Figure 13.2
Transistors.

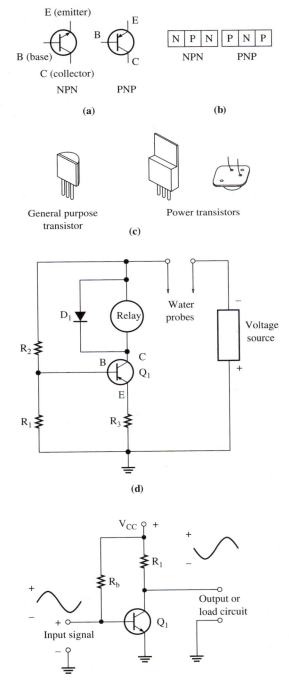

E (emitter)

B (base)

C (collector)

NPN

B

E

C

PNP

(a)

| N | P | N | P | N | P |

NPN   PNP

(b)

General purpose
transistor

Power transistors

(c)

D_1 Relay   Water
probes

R_2

R_1

B   C
Q_1
E

R_3

Voltage
source

(d)

V_{CC} +

R_1

R_b

Input signal

+

−

Output or
load circuit

Q_1

(e)

259

Silicon Controlled Rectifiers (SCRs)

The SCR is a three-terminal (cathode, anode, gate) solid-state latching switch. In its normal state it blocks current flow from cathode to anode. However, the component will conduct in the forward direction when a positive voltage is applied to the gate electrode. Furthermore, once conduction takes place, it continues even when the control voltage to the gate is removed. The SCR ceases conducting only when the anode voltage is removed, reduced, or reversed.

In Figure 13.3a we see the schematic symbol and characteristic curve of a typical SCR. In Figure 13.3b we have an SCR test circuit for the 1 A/200 V C106B1 SCR. With S_1 closed, the SCR is in the "ready" state. When S_2 closes, the SCR's gate receives a positive voltage through R_1. The SCR now conducts, current flows from its cathode to anode (electron current flow), and the load is activated. Even if S_2 is opened, the SCR will continue to conduct. If S_1 is opened, however, the anode voltage is removed, and the SCR ceases to conduct. If S_1 is again closed but S_2 remains open, the SCR will stay off. Simply closing S_2 will turn it on again.

SCRs are rated in current- and voltage-handling ability. For example, a 1 A/200 V SCR will conduct *up to* 1 ampere while handling *up to* 200 volts. Will such an SCR work in a circuit delivering 500 milliampere at 6 volts? You bet.

SCRs are used extensively in burglar alarms, color organs, strobe lights, and industrial applications—anywhere a solid-state latch is required.

Speaking of burglar alarms: The circuit in Figure 13.3b can easily be converted to a low-cost protection device. Simply replace S_2 with a normally-open contact, S_1 with a key switch, and provide a buzzer for the load. When the N.O. contact closes, even momentarily, the buzzer sounds. And it will keep on squawking until S_1 is opened, or the battery runs down.

Want an easy, cheap one-channel color organ that will flash a string of Christmas lights to the sound of music? The circuit in Figure 13.3c, with an SCR as its core component, will do the trick. During the positive half-cycle of the incoming 60 Hz ac, the SCR is in the "ready" state. Any audio signal tapped off of R2 and fed to the SCR's gate will trigger it into conduction, turning on the string of lights. When the 60 Hz ac waveform goes negative, the SCR turns off (its anode voltage is reversed). Thus, 60 times a second the SCR responds to an audio signal from a stereo, tape player, or compact disc.

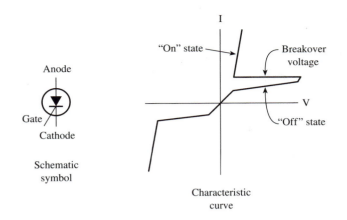

(a)

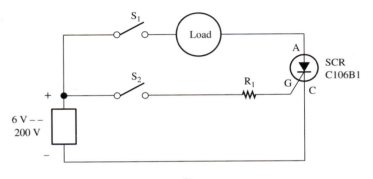

(b)

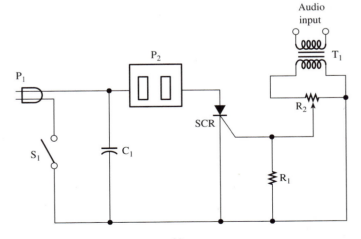

(c)

Figure 13.3
Silicon controlled rectifiers (SCRs).

261

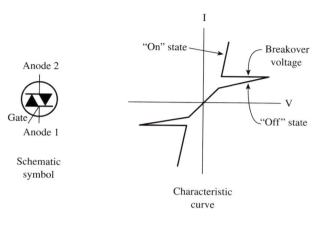

(a)

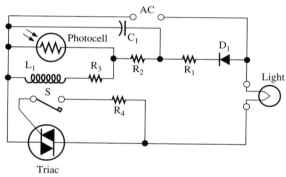

(b)

Figure 13.4
Triacs.

Triacs

A triac is a back-to-back SCR. It functions as an electrically controlled switch for ac loads. The schematic symbol and characteristic curve for a typical triac are shown in Figure 13.4a.

The triac has an NPNPN structure, which allows it to be triggered into either forward or reverse conduction by a pulse applied to its gate electrode. Thus a triac, unlike an SCR, will pass alternating current.

Like an SCR, however, the triac is rated in terms of current- and voltage-handling ability. The former ranges from 1 ampere to over 40 amperes; the latter, from 200 to 600 volts.

Triacs are used to control high-current, high-voltage ac loads. In dc circuits, we use the SCR; in ac circuits, where current and voltage tend to be higher, the triac is the choice.

One application for a triac is in the automatic night light, a schematic of which is shown in Figure 13.4b. When no light falls on the photocell, its resistance is high and most of the current flows through L_1, the coil of a reed switch. With current through L_1, contact S closes, and the triac fires, causing the light to glow. When light is shining on the photocell, its resistance decreases, causing most of the current to pass through the photocell. The reduction in the current through L_1 opens the reed switch contacts and the triac turns off the light.

Voltage Regulators

A voltage regulator is a circuit that holds an output voltage at a predetermined value regardless of normal input voltage changes or changes in the load impedance. There are fixed and variable voltage regulators. The former have a fixed output, such as 5.0, 6.0, 12.0, and 18.0 volts. The latter will hold their output voltage steady within a predetermined range. For example, the LM317 is a 1.2- to 17-volt regulator. When set to any voltage within that range (usually with a potentiometer), it will hold the output at the selected voltage.

Many voltage regulators come in a three-terminal package. They have only one input terminal, one output terminal, and one ground terminal, as shown in Figure 13.5a. Such regulators are chosen primarily to satisfy the voltage and current requirements of the load. To illustrate, the LM7805 is a fixed 5 V/1 A voltage regulator. It will hold the output voltage at 5 volts while the load draws up to 1 ampere. If the load exceeds 1 ampere,

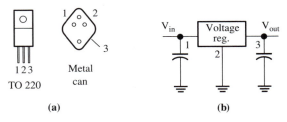

Figure 13.5
Voltage regulators.

the voltage regulator will automatically shut down. It will resume operation when the load requirement is reduced below 1 ampere.

Voltage regulators are a "must" for dc power supplies. Considering their simplicity and low cost (in many cases, under $1), there is no reason why they shouldn't be designed into every power-supply project (Figure 13.5b).

Optical Couplers (Isolators)

An optical coupler (also known as an optical isolator) is a component that couples signals from one electronic circuit to another by means of infrared radiation or visible light. Thus complete electrical isolation from the primary and secondary circuit is achieved. Figure 13.6a shows a typical optical coupler consisting of an LED to convert the electrical signal of the primary circuit into light, and a phototransistor in the secondary circuit to reconvert the light back into an electrical signal. There are other possible light receptors, such as SCRs, triacs, Darlington amplifiers, photodiodes, and photoresistors.

Figure 13.6
Optical couplers (isolators).

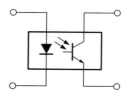

Typical optical coupler

(a)

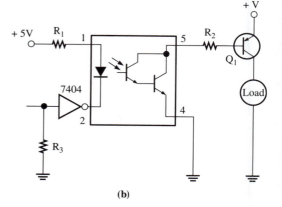

(b)

Optical couplers are used anywhere it is desirable to completely isolate one circuit from another. Ratings depend on the current- and voltage-handling ability of the light source as well as that of the light-converting element.

In a typical application, we see a PS2002B optical coupler, shown in Figure 13.6b, being used to provide isolation. When the LED turns on, so does the Darlington transistor; its output at pin 5 is thus brought to ground. When the LED is off, the output at pin 5 is in the high-impedance state. The output of the Darlington transistor is in turn fed to a general-purpose transistor that switches on a relay, triac, or similar device. It's the latter component that actually controls a high-voltage (120V) load.

There you have it—an introduction to six solid-state electronic control components. As you pursue your study of electronics, you'll take courses that expand on what we've covered here. Your introduction will grow to in-depth knowledge.

Summary

In Chapter 13, we examined six solid-state components. We looked at the diode, a component that allows current to flow in one direction only. We saw how the transistor can both switch and amplify current. We explored the SCR, a solid-state latching switch. And we looked at the SCR's big brother, the triac. We investigated the workings of a voltage regulator, seeing how it holds a power supply's output voltage steady. Finally, we examined the optical coupler, a tiny component that isolates one circuit from another.

Review Questions

1. A _____ is a two-element component that allows current to flow through it in one direction only.

2. A bipolar transistor is a semiconductor component that can both _____ and _____.

3. There are two broad types of transistors, classified according to _____-handling ability.

4. An SCR is a solid-state _____ switch.

5. An SCR must have a _____ voltage on its anode and a _____ voltage on its cathode in order to be in its ready state.

6. A triac is a back-to-back _____.

7. A voltage regulator is a circuit that holds an output voltage at a predetermined value regardless of normal input voltage changes or changes in the _____ impedance.

8. Voltage regulators are a "must" for dc _____.

9. Optical couplers are used anywhere it is desirable to completely _____ one circuit from another.

10. Optical couplers are also known as optical _____.

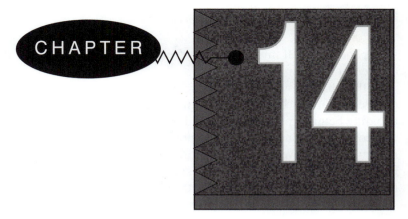

CHAPTER

14

Electronics Safety:
Where the Danger Lies

Objectives

In this chapter you will learn:

- How your eyes, nose, throat, and skin can be damaged in a laboratory.
- How electrical shock can kill.
- How much current is enough to kill.
- About the *do*s and *don't*s of working with electricity.
- About the *do*s and *don't*s of handling toxic and hazardous chemicals.
- About safety hazards caused by the misuse of hand tools.
- About safety hazards caused by the misuse of power tools.
- How to set up a safe working environment at home or in the lab.

As an electronics enthusiast, you need to be aware that the electricity pulsing through your projects and equipment could, if not properly contained, kill you. Furthermore, the myriad of toxic and hazardous chemicals you're using to clean, lubricate, and etch circuit boards are often noxious and corrosive. And then there are the hand and power tools you use to cut, drill, grind, and solder. If mishandled, you're likely to wind up laid up, at home or even in the hospital.

In this chapter, we'll look at external body parts susceptible to damage; electrocution, the heart of the matter; electricity, the tingle that hurts; avoiding being exposed to toxic and hazardous chemicals; safety hazards caused by the misuse of hand and power tools; and creating a safe working environment. We will do so, for the most part, in a *dos*-and-*don'ts* format that will provide you with the specifics needed to stay safe and keep building, testing, and repairing electronic projects and devices as you pursue your study of electronics.

External Body Parts Susceptible to Damage

The human body is a fragile thing. It's easily cut, bruised, burned, irritated, shocked, and poisoned. External organs associated with our five senses and internal organs vital to our very lives are susceptible to injury, not only in the workplace, but at home in the den, garage, or basement.

Injury to the *eyes* can cause permanent blindness. Small objects, such as clipped component leads and flying solder balls, can stab and burn sensitive parts of the eye. Furthermore, a splash of acid or the spray from a cleaning solvent can, at the very least, irritate; at the worst, it can destroy the entire eye. Even a bright flash of light from a strobe light project, for example, can cause a serious accident. If you are only momentarily blinded, it may be enough to make you susceptible to other forms of injury.

Loud sounds and high-pitched tones can be very irritating to the *ears.* In extreme cases they may cause temporary or even permanent hearing loss. It isn't just heavy metal rock bands that can produce such noises. Many types of electronic devices can blast forth with ear-piercing decibels.

You may not think your *nose* is vulnerable to dangers found in your work area. Yet toxic fumes can be irritating to the membranes lining the nasal passages. More importantly, the nose is one entry leading to the lungs. Damage there can be very serious, as it may affect your ability to breathe properly.

Your mouth, or *throat,* can also be an opening for foreign and hazardous particles. Of course, you wouldn't knowingly put any acid solution or solder wire (which is 40 percent lead) in your mouth. But if etching acid (ammonium persulfate, for example) gets under your fingernails, and you bite your nails, you could be in trouble. The greatest danger, however, lies in using your mouth as a temporary vise to hold small pieces of hardware, tiny transistors, and short strips of wire. Accidental swallowing, or a quick slap on the back from a fellow student, and presto—it's time to have your stomach pumped.

Then there is the *skin,* the outer layer designed to protect the entire body. Destruction of nerve endings as a result of electrical shock, burns from soldering irons and hot components, acid irritations, and, of course, common cuts, bruises, and abrasions can occur in an environment where safety is lax.

Clearly, your entire body is subject to harm. However, there is no cause for alarm, only a healthy concern. Shortly we'll see how you can eliminate most of these dangers from your work area by taking simple precautions and following commonly accepted safety practices.

Electrocution: The Heart of the Matter

Protecting external body parts is one thing; preventing damage to various internal organs, particularly the *heart,* is vital. If the heart stops, even for only 4 to 6 minutes, death will result. Many factors can cause the heart to stop beating. For purposes of discussion here, **electrocution,** or **electrical shock,** is the most relevant. Let's see just what electrical shock is and how it can destroy the heart.

Electrical shock is the passage of current through the body. Current flows through the human body, as it does anywhere else, when a complete circuit exists. Such a circuit is shown in Figure 14.1a. A source of voltage and a conducting path (shown in the figure as having a given amount of resistance) are all that's required. Note, however, the path for current flow must be complete. In Figure 14.1b, no current flows to point A. The reason? Although it has a place to come from (in this case, the battery's negative terminal), it has no place to go to (in this case, the battery's positive terminal).

The same is true when the human body is the conducting path. As illustrated in Figure 14.1c, even though the person is touching the positive terminal of the voltage source, he's in no danger. As long as he's wearing insulated boots, current can find no path through him to the negative, or ground, terminal of the battery. However, if he completes the circuit by, for instance, touching the negative battery terminal with the other hand (possibly through a ground connection), current will indeed flow, as shown in Figure 14.1d. Just how much current depends on a number of factors.

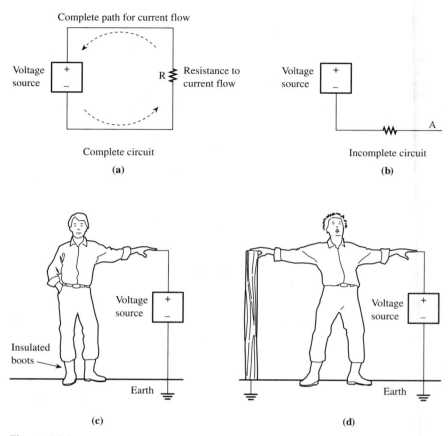

Figure 14.1
The human body as a conducting path.
Source: From *Electronic Project Design and Fabrication,* Fifth Edition (p. 31), by R. A. Reis, 2002. Upper Saddle River, NJ: Prentice Hall. Copyright 2002 by Prentice Hall Publishing Company. Reprinted by permission.

One way to increase current flow is to reduce circuit resistance. Body re-sistance may be quite high if skin moisture is low and there are not cuts or abrasions at the point of electrical contact. In such cases, little current will flow and only a mild shock may result. Nonetheless, if any of these factors are reversed, resistance will be lowered and large amounts of current could result. If the path for current flow is through the chest, the heart can receive a lethal dose of current, an electrical shock. The heart will then most likely go into fibrillation (rapid, irregular muscle contraction) and stop beating.

How much current is enough to kill? Although the amount varies widely and especially depends on the organs current passes through, it's

Figure 14.2
How much current is enough to kill?
Source: From *Electronic Project Design and Fabrication,* Fifth Edition (p. 32), by R. A. Reis, 2002. Upper Saddle River, NJ: Prentice Hall. Copyright 2002 by Prentice Hall Publishing Company. Reprinted by permission.

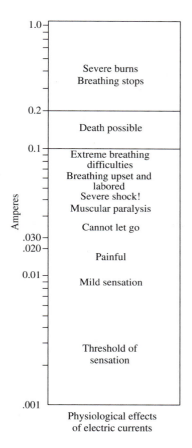

Physiological effects
of electric currents

less than you probably think. Figure 14.2 gives the effects of even mild doses of current. A mere 1 to 20 milliampere can cause a painful sensation; at 30 milliampere, breathing may stop; and 100 to 300 milliampere is enough to cause electrocution. Clearly, even small amounts of unwanted current through the body can be very dangerous to your health.

Electricity: The Tingle that Hurts

There is good news and bad news. The good news is that today's electronic devices, many of which are solid-state and battery powered, are much safer than their predecessors of the vacuum tube era. Instead of circuits with voltages in the hundreds and currents in amperes, we now have equipment operating on 5 to 12 volts dc and drawing currents of less than

50 to 100 milliampere. The bad news is that many systems, even though employing solid-state circuitry, do not get their power directly from a battery but rather from the ac line. And any electronic project plugged into the wall outlet is a potential death trap.

The problem with ac is twofold. The ordinary wall outlet will supply up to 25 amperes of current, certainly enough under many conditions to kill you. In addition, ac tends to "hang on." The victim can be prevented from releasing the source of voltage, thus increasing the damage to the body.

But circuits powered from the ac line are not the only problem. A serious shock can be had from many circuits, dc or ac, containing charged capacitors, faulty wiring, or shorted components. Treat every electronic circuit as potentially hazardous. Electricity, though not to be unduly feared, must be respected.

One solution to being safe while working with electricity is to follow some basic "*do*s and *don't*s." Here are the more important ones:

The *Do*s

- *Do* work with one hand behind your back while testing live circuits. In that way if you complete the path for current flow, at least it won't be through your heart.

- *Do* use an isolation transformer while working on ac-powered equipment. This device isolates the powered equipment from the power source, adding a strong measure of safety.

- *Do* use 3-conductor grounded line cords and polarized plugs with ac-operated equipment. Both items reduce the danger from short-circuit chassis.

The *Don't*s

- *Don't* install or remove any electronic components while the circuit is connected to a power source. Following this procedure will protect you as well as the component.

- *Don't* overfuse. Using a fuse with a higher rating than is recommended only defeats the fuse's purpose.

- *Don't* cut wires carrying electricity. Again, it isn't just the ac line cord that is lethal. Assume all wires carry enough current to harm you.

There is not much you can do for yourself if severe electrical shock occurs. All the more reason you should avoid working alone whenever possible. At the very least, if you are fiddling around in a garage or basement, make sure someone is in the house and knows what you are doing.

When I am out in the garage alone, I plug in (don't laugh) a line-operated "nursery monitor," the kind used to listen to your baby from another room. The transmitter is in the garage; the receiver, in the house with my wife. If anything happens, chances are she'll hear me yelling.

Exposure to Toxic and Hazardous Chemicals

With toxic and hazardous chemicals, you should be concerned about inhaling vapors, swallowing liquids, acid burns on the skin, contact with the eyes, and the overall danger of fire and explosion. Chemicals such as etching solutions, spray paints, component cleaners, glues, photographic developing solutions, and general household solvents can cause health hazards if improperly used or stored. To minimize risks, follow the *do*s and *don'ts* listed here:

The *Do*s

- *Do* read the labels on all the chemicals you use. Pay particular attention to printed warnings.
- *Do* wear eye protection when working around hazardous chemicals.
- *Do* wear rubber gloves when working with acid solutions. Be sure they are washed or thrown away when work is done.
- *Do* work in well-ventilated areas. This is particularly important when using paint and chemical sprays.

The *Don'ts*

- *Don't* pour solvents, paints, or acids down the drain. In addition to being a health hazard, in many localities it's illegal.
- *Don't* transfer chemicals to unlabeled containers.
- *Don't* inhale fumes from soldering paste. Such fumes can be harmful to your lungs.

If an emergency occurs, follow all first-aid directions printed on the label of the chemical you have been using. Read the label **before** use so you won't be confused when trouble develops. For example, some instructions tell you to induce vomiting if the substance is swallowed; others tell you not to. Do what the label says and call a physician immediately.

More specifically, in the case of acid burns and eye irritations, wash and flush thoroughly with cold running water. When exposure to vapors has occurred, get yourself outside in the open air.

Safety Hazards Caused by the Misuse of Hand Tools

Hand tools, those held in the hand and not powered by electricity, can be more dangerous than power tools. Why? Because they are used more often and we tend to dismiss their risk. Do not be deceived. A slip of a hacksaw blade or the launching of a loose hammerhead can maim you—or someone else—for life. Whether it's a puncture from a file tang, a gash from a saw blade, or a slit from a knife, hand tools can do plenty of damage. To avoid the obvious dangers, consider this list of *do*s and *don't*s:

The *Do*s

- *Do* keep hand tools sharp. A dull tool is more dangerous than a sharp one because with a dull tool you tend to apply more pressure, increasing the likelihood of slippage.
- *Do* secure all small pieces of work in a vise or with appropriate clamps.
- *Do* put sharp tools away when not in use. Leaving knives, awls, and punches lying around for someone to lean on is extremely dangerous.

The *Don't*s

- *Don't* use a file without a handle. An exposed file tang can easily puncture your hand or wrist.
- *Don't* carry hand tools in your pocket; they may injure you or someone else.
- *Don't* run your hands over the edge of sheet metal. Such edges are sharp and can cause severe cuts.

Injury from hand tools usually results in a cut or burn. What specific action to take when a laceration occurs depends on its severity. In all cases, clean the wound, apply direct pressure, and elevate the injured body part. With regard to burns, flush the affected area with cold water (do not apply ice directly) and cover with a clean cloth. Seek medical aid.

Safety Hazards Caused by the Misuse of Power Tools

Power tools, those that use electricity, can burn, cut, scrape, and even hit you with flying objects. Because most of them plug directly into the wall outlet, they can also give you a nasty electrical shock. Injuries from power tools occur when long hair or unsuitable clothing get caught in a revolving machine, when small objects fly out of a drill press vise, when guards or safety devices are removed, or when a hot soldering iron is left where someone can lean against it. To prevent such injuries, follow the *do*s and *don't*s listed here:

The *Do*s

- *Do* turn on and off all power tools yourself. Do not allow others to do it for you.
- *Do* make sure all objects being drilled or cut are securely fastened. Small objects, and especially sheet metal, should be held in a vise, clamp, or suitable gripping tool such as pliers or a vise grip.
- *Do* remove the chuck key from a drill press before turning on the machine. A flying chuck key is a frequent cause of injury in the laboratory.
- *Do* grasp a soldering iron only by the handle. Don't reach for a falling soldering iron—let it fall.

The *Don't*s

- *Don't* ever leave power tools unattended. If the tool is on, stay with it until it's off—and until all moving parts have stopped moving.
- *Don't* wear clothing or accessories (jewelry) that can get tangled in revolving machine parts.
- *Don't* stand in the direct "throw" of any machine. Don't line yourself up with a revolving saw blade or spinning grinding wheel.
- *Don't* remove guards or safety devices from power tools. They have been installed for a good reason; keep them in place.

Injuries resulting from misuse of power tools are similar to those occurring with hand tools, so the first-aid procedures are essentially the same. The accident itself, however, may be more severe. Get help immediately, then follow the appropriate first-aid measures.

A Safe Working Environment

Following the *do*s and avoiding the *don't*s presented here will do much to keep you safe while working in electronics. Yet, more than anything else, establishing a good, clean, well lit working environment is the key to safety success.

If you're working at home, try to locate a workbench in an area that will provide a minimum of distractions. Such a location will increase the likelihood the bench will be used only for its intended purpose, electronic project building and circuit testing and repair.

The workbench should remain relatively clean. The dirt and mess associated with normal mechanical assembly (such as would occur when working on your car, for example) is incompatible with modern electronics fabrication and repair.

Your workbench should be bathed in direct and indirect light. You might not think this lighting is necessary, but when you are working with tiny circuits and minute letters and numbers printed on small components, you'll need all the brightness you can muster. A drafting lamp can provide a concentrated direct beam of light. An even better choice is a swivel lamp with an illuminated magnifier. Such lamps are particularly useful when you're inspecting or assembling printed circuit boards.

That should do it. Enjoy your electronics experimenting, building, and repairing. And stay safe.

Summary

In Chapter 14, we looked at safety, at home and in the lab. We saw how injury to your eyes, nose, throat, and skin can occur when working around electricity. We looked at electrocution and how little current it takes to kill. We examined *do*s and *don't*s with regard to electricity, exposure to toxic and hazardous chemicals, and misuse of hand and power tools. We concluded with an examination of what it takes to create a safe working environment.

Review Questions

1. Small _____, such as clipped component leads and flying solder balls, can stab and burn sensitive parts of the eye.

2. Electrical _____ is the passage of current through the body.

3. The ordinary wall socket will supply up to _____ amperes of current.

4. Do work with one hand behind your back while testing _____ circuits.

5. Don't pour solvents, paints, or acids down the _____.

6. Do secure all small pieces of work in a vise or with appropriate _____.

7. Do make sure all objects drilled or cut are _____ fastened.

8. Don't stand in the direct "_____" of any machine.

9. Don't remove guards or safety devices from _____ tools.

10. A workbench should be bathed in direct and indirect _____.

CHAPTER

Getting Started with SMT:
Surface Mount Technology for the Electronics Experimenter

Objectives

In this chapter you will learn:

- About the packaging revolution called surface mount technology (SMT).
- How surface mount components (SMCs) are configured.
- What tools and materials are needed to work with SMCs.
- How to attach SMCs to a PC board prior to soldering them in place.
- How to solder SMCs to a circuit board.
- How to remove SMCs from a circuit board.

Surface mount technology (SMT) is a packaging revolution that involves attaching tiny, essentially "leadless" components to pads *on top of* a printed circuit board—on its surface. Thus the name, Surface Mount

Technology. SMT is contrasted with traditional insertion mount technology (IMT), which uses components that do have leads, leads that are inserted *through* the PC board.

The SMT approach results in significant advantages, such as circuit board size and weight reduction, enhanced circuit performance and reliability, improved manufacturing of stuffed circuit boards (they all involve automated assembly), considerable cost reduction, and development of new products dependent on SMT: the credit card–size medical records disk, for example. As a consequence, today 90 percent of electronic assemblies are surface mount. Every electronics technician needs to know about SMT.

In this chapter, we see how to handle tiny surface mount components. We examine their basic characteristics, the tools and methods needed to work with them, and how to put SMCs in their place. By the time you've completed Chapter 15, you'll know your way in the SMT world.

Teeny, Tiny SMCs

SMT circuit assembly is an automated process. Surface mount components are designed to be packaged in reel dispensers, put in place by X-Y pick-and-place machines, and soldered onto a PC board using flow (wave) or reflow methods. In an automated SMT assembly line, no human hands ever touch the components or the printed circuit boards.

Nonetheless, while mass-produced SMT circuits are machine assembled, prototypes and SMT projects are put together using hand tools. True, special tools and equipment are available for such tasks, and they should be used whenever possible. However, much can be done with traditional tools, used in somewhat nontraditional ways.

Before we see how to put SMCs in their place, we need to know more about them. We look first at discrete passive components, such as resistors and capacitors; then at discrete active components, diodes and transistors; at integrated circuits; and, as a catch-all, at a few SMCs that don't fall into any other categories, trimmers and LEDs, for example.

The *leadless chip resistor* is the most popular discrete passive surface mount component. See Figure 15.1a. Available in standard values from 10 ohms to 10 megohms, and in wattage ratings as low as ⅙₆, and as high as ¼, watt, SMC resistors are a universal surface mount component.

The *ceramic chip capacitor* is also a widely used SMC. See Figure 15.1b. Available in both nonpolarized and polarized versions, the former range in value from 1 pF to 1 μF; the latter (most often tantalums), from 0.1 μF to 100 μF. Voltages go as high as 1000 volts.

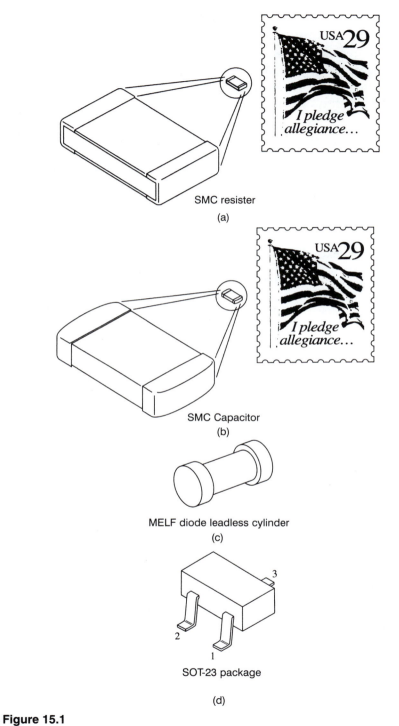

SMC resister

(a)

SMC Capacitor

(b)

MELF diode leadless cylinder

(c)

3

2

1

SOT-23 package

(d)

Figure 15.1
SMC discrete passive and active components.

281

Of the discrete active SMCs, the *two-terminal diode* in a MELF (metal-electrode face) leadless cylinder is popular. See Figure 15.1c. The package, also known as an SOD, for small outline diode, is ¹⁄₁₀″ in diameter and ⅕″ long. Diode arrays are available, too. See Figure 15.1d.

Low-power *transistors* come in SOT-23 (small outline transistor) packages, as shown in Figure 15.2a. The device is only 0.118″ long, 0.05″ wide, and 0.04″ tall. Note the **gull-wing** lead configuration. Transistors required to dissipate more power than the general purpose type appear in the SOT-89 package, shown in Figure 15.2b. The four-terminal SOT-143 is also a popular SMC case style (Figure 15.2c). Measuring 0.118″ long, 0.05″ wide, and 0.04″ tall, it is used for bridge rectifiers, field-effect transistors, and diode pairs. Note one lead is wider than the others to provide a convenient index mark.

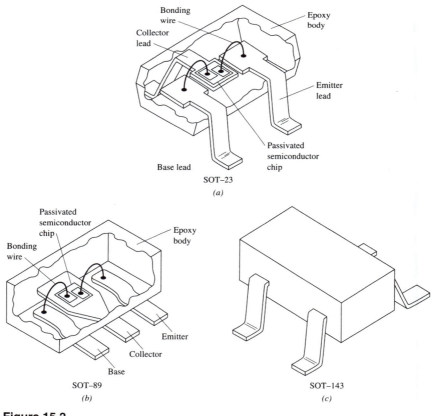

Figure 15.2
SMC transistors.

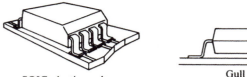

SOIC plastic package   Gull wing

Figure 15.3
Small outline integrated circuit (SOIC).

Surface mount *integrated circuits* are available with a variety of lead configurations. The type you're most likely to encounter, however, is the **small outline integrated circuit (SOIC)** shown in Figure 15.3. Note, again, the gull-wing lead configuration. While the SOIC ICs look like miniature DIPs (traditional through-hole ICs), they're one quarter the size of their bigger cousins. Furthermore, lead spacing is 0.050″, half that of the traditional DIP.

Many other electronic components come in surface mount configurations. While we can't discuss them all here, two deserve note. Surface mount *trimmer resistors* and *capacitors,* as seen in Figure 15.4a, are common. Typical potentiometer values range from 100 ohms to 2 megohms. Trimmer capacitors range in value from 1.4 pF to 50 pF.

Subminiature surface mounted LEDs, (Figure 15.4b), find wide application. Note, once more, the gull-wing lead configuration.

Tools and Materials

Before we examine three methods used to hold SMCs in place while hand-soldering with solder wire, look at Table 15.1. If you plan to assemble SMCs, you'll want to obtain the tools and materials listed.

The soldering iron, 25 to 40 watts, should have a tinned tip with a conical shape, ¹⁄₁₆″ in diameter. A tweezer, with forceps-style tips, is mandatory for picking up tiny SMCs. A vise is useful for securing the PC board while components are being attached. If you don't have a desktop magnifying glass, get one; or, at the least, purchase an inexpensive hand-held unit.

The materials listed in Table 15.1 are self-explanatory. Solder should be 63/37 or, better yet, silver-bearing solder at 62/36/2. While a 0.020″ diameter is OK, 0.015″ diameter is better. A noncorrosive liquid flux and drop dispenser are necessary. So is a light-duty spray defluxer. A general-purpose plastic cement, the kind for plastic, wood, or metal, is fine. When

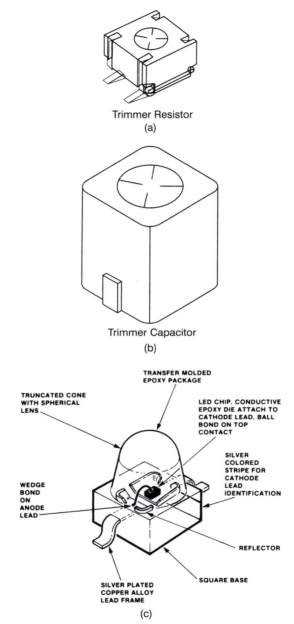

Trimmer Resistor
(a)

Trimmer Capacitor
(b)

TRANSFER MOLDED
EPOXY PACKAGE

TRUNCATED CONE
WITH SPHERICAL
LENS

LED CHIP, CONDUCTIVE
EPOXY DIE ATTACH TO
CATHODE LEAD, BALL
BOND ON TOP
CONTACT

SILVER
COLORED
STRIPE FOR
CATHODE
LEAD
IDENTIFICATION

WEDGE
BOND
ON
ANODE
LEAD

REFLECTOR

SILVER PLATED
COPPER ALLOY
LEAD FRAME

SQUARE BASE

(c)

Figure 15.4
Other SMCs.

284

Table 15.1
Tools and materials for
attaching SMCs.

Tools
• Soldering iron
• Tweezers
• Vise
• Magnifying glass (hand-held)
Materials
• Solder
• Liquid flux
• Drop dispenser
• Defluxer
• Adhesive (plastic cement)
• Solder wick
• Clear tape

choosing solder wick, select 0.030″ diameter. Ordinary clear tape ½″ in diameter works. Any other materials you need are probably lying around your house.

Putting SMCs in Their Place

Three ways exist to hold SMCs down in preparation for soldering. You can tape, stick, or tag them to the PC board.

With taping, grasp the clear tape and "touch" it to the SMC, thus picking up the latter. See Figure 15.5a. Then tape the SMC onto its PC board pads. With the component now held in place, solder the exposed terminal (Figure 15.5b). Remove the tape, then solder the remaining terminal(s) to their pads (Figure 15.5c).

Gluing is another way to hold an SMC to the circuit board in preparation for soldering. A general-purpose glue (in a jar), such as the General Purpose Plastic Cement, 10-324, from GC Electronics, works well.

The best way to dispense glue is with a toothpick. Dip the pick into the liquid, stopping just as it touches the surface. Then place a "dot" of glue on the PC board, being careful not to get any on the copper pads. (See Figure 15.6.) Now, gently press the component into place and let the glue dry for about 30 minutes.

While many use the tape and stick methods just described, most prefer the tag soldering approach. It's easy, quick, and involves no additional materials such as tape or adhesives. All you need is solder wire and a soldering iron.

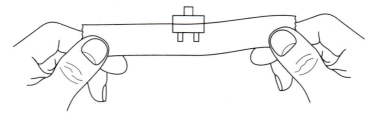

(a)

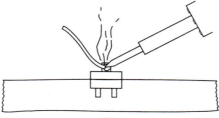

(b)

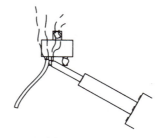

(c)

Figure 15.5
Taping SMCs down.

Figure 15.6
Dispensing adhesive.

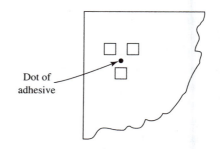

Dot of
adhesive

286

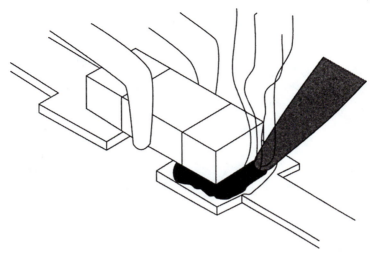

Figure 15.7
Soldering two-terminal SMCs to a PC board.

First, dab liquid flux onto *one* PC board pad. Then melt solder on top of the pad. Allow the solder to solidify.

Next, with the SMC held in a tweezer, rest the component on its PC board pads and hold in place. Using a soldering iron held in your other hand, reflow the solder, causing the component terminal to "sink" close to the board surface. See Figure 15.7. Remove the iron, allow the solder to again cool, and release the tweezer. Now with the component "held" in place, solder the other terminals or pins in a traditional manner.

The Solder Connection

Once an SMC is held down, it's time to solder the remaining terminals or leads in place. For two-terminal discrete SMCs, begin cleaning the PC board pads with steel wool. Next, using the tag method, since it will give you practice in soldering, solder one component terminal to its pre-tinned pad as described earlier. Now solder the other terminal as you would any traditional joint, though apply just a dab of solder.

To solder gull-wing leads on three- and four-lead components, proceed as above, only be sure to apply plenty of liquid flux before soldering. Doing so makes for a much more effective solder connection. See Figure 15.8.

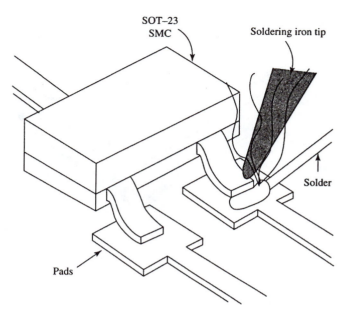

Figure 15.8
Soldering gull-wing leads.

Hand soldering DIP gull-wing SOICs to a circuit board is not difficult. Just follow these ten steps:

1. Make sure pads are clean and ready for soldering.
2. Create a pool of solder on a corner pad in preparation for tag soldering.
3. Hold the IC with a tweezer or your fingers.
4. Place the IC onto the copper pads. Be sure to center the IC right-to-left as well as top-to-bottom. Hold the IC in place.
5. Apply the soldering iron tip to the pretinned pad. Hold the iron in place just long enough to flow the solder.
6. Pivot the IC about the soldered pin as necessary to again align all pins with their respective pads.
7. Solder a second pin diagonal to the first. In doing so, apply a small amount of liquid flux.
8. Bathe one line of pins with liquid flux.
9. Solder the line of pins, one at a time. Move quickly, yet adhere to the four traditional soldering steps: apply heat; apply solder;

remove solder; remove heat. Apply liquid flux to the opposite
row of pins and solder as just described.

10. Inspect your work under a magnifying glass.

Desoldering

There will be times when you'll need to remove a component (wrong
component, component in backwards, component severely misaligned,
etc.). As a general procedure, you simply dip solder wick into liquid flux,
place it over the lead or pin, and apply heat with a soldering iron. When
all possible solder has been removed, the component should lift up with a
slight pull.

Summary

In Chapter 15, we learned why surface mount technology is a packaging
revolution sweeping all electronics. We saw how tiny surface mount components are configured, from leadless chip resistors to gull-wing integrated circuits. We examined hand tools and materials necessary to build
SMT prototypes and projects. And we explored three methods of attaching SMCs to a PC board prior to soldering them in place. We concluded
by looking at ways to both solder SMCs to a PC board and remove them
when necessary.

Review Questions

1. The SMT approach results in significant advantages, such as circuit board _____ and _____ reduction.

2. SMT circuit assembly is an _____ process.

3. The _____ _____ _____ is the most popular discrete passive surface mount component.

4. SOIC stands for _____ _____ _____ _____.

5. When working with SMCs, a forceps-style _____ is necessary.

6. When working with SMCs, solder _____″ in diameter is desirable.

7. With the taping method of temporarily holding SMCs in place, _____ tape should be used.

8. The best way to dispense glue is with a _____.

9. The tag solder method of holding SMCs in place is easy, quick, and involves no additional _____ such as tape or adhesives.

10. When soldering SMCs in place, use plenty of liquid _____.

Virtual Electronics

Objectives

In this brief chapter, we will:

- Examine the elements of circuit design, simulation, and analysis.
- Look at a typical software package, Multisim V6.2, from Electronics Workbench.

It is now possible to both design a circuit and simulate its operation on a computer. Actually, we've been doing so for some time. In the 1970s, powerful minicomputer systems, housed in large engineering laboratories, did just that. What's new, of course, is the ability to do such design and simulation on a personal computer, using cost-effective software selling for less than $400. There is little reason, therefore, or excuse, to jump from design sketch to breadboarding layout. Today, you can verify a design before breadboarding with computer-based design and analysis tools.

The Virtual Circuit

The virtual circuit is here. A circuit that appears to be, rather than actually is, can be represented on a computer screen. First, the **design,** in schematic form, is created. Next, the circuit is **simulated, run,** with LEDs flashing, speakers sounding, motors turning, and instruments displaying.

Then, an **analysis** of its operation, using analysis graphs, is made. Thus the design, simulation, and analysis of a virtual circuit is complete. It's not the real thing, of course, but the virtual circuit can get you closer to reality while saving considerable laboratory time.

Circuit Design

A circuit, real or virtual, consists of electronic components connected by wires (or traces on a PC board). To design a circuit on a computer, you create a schematic consisting of components wired together.

Your components cannot just be drawn on the screen, however. First they must be endowed with characteristics in the form of values and models. Then, when they are wired together and the resulting circuit is simulated, the circuit "works." If all you do is draw the components, as is done in a typical CAD package, you have only schematic symbols. Such symbols lack "intelligence": no circuit simulation or analysis is possible.

All circuit analysis programs have libraries chock full of components to be called up and placed in the drawing area. These components are predefined, having mathematical properties that allow them to function when the circuit is enabled.

The components, of course, can be moved, rotated, flipped, and copied. They can be assigned labels, values, and reference designations, and, as you would expect, they can be eliminated (see Figure 16.1).

Once components are placed on the computer screen, you wire them together by drawing lines from one terminal to another. It is simply a matter of pointing to a component terminal with the mouse and then dragging the mouse to a terminal on another component (see Figure 16.2.) The wire is automatically routed at right angles, without overlapping other components.

With the circuit wired, your design is complete. Now comes the simulation to see if it works.

Figure 16.1
Rotating and flipping components.

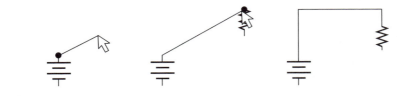

Figure 16.2
Wiring Components.

Circuit Simulation

The circuit simulator calculates a numerical solution to a mathematical representation of a circuit you create. As mentioned earlier, for calculating to occur, each circuit component must be represented by a mathematical model. Mathematical models link the schematic in the circuit window with the mathematical representation for simulation. The accuracy of component models determines the degree to which simulation results match real-world circuit performance.

Most general-purpose simulators have four main stages: **input, setup, analysis,** and **output.**

1. At the input stage, the simulator reads information about your circuit.
2. At the setup stage, the simulator constructs and checks a set of data structures, containing a complete description of your circuit.
3. At the analysis stage, which is the core of the simulation, circuit analysis specified in the input stage is performed.
4. At the output stage, you view simulation results on instruments or on graphs appearing when you run an analysis.

Is simulation worthwhile? Do simulators work? Well, ask any engineer who worked in the historic Pathfinder Mission to Mars. Without computer simulation, there wouldn't have been an Independence Day landing on the planet's surface in 1997.

When it comes to circuit design, too, simulation is valuable. True, it's no substitute for the real thing. Nonetheless, running a simulation allows you to get your prototype stage more quickly.

Circuit Analysis

Once your circuit is designed and operating, you can do an analysis on it and find out what's really going on. There are many types of analyses: dc operating point analysis, ac frequency analysis, transient analysis, Fourier analysis, noise analysis, distortion analysis, and parameter sweep analysis. The first three warrant our brief attention.

DC operating point analysis determines the dc operating point of a circuit. The results are displayed on a chart that appears when the analysis has finished. The chart lists node dc voltages and branch currents.

In *ac frequency analysis,* the dc operating point is first calculated to obtain linear, small-signal models for all nonlinear components. The result is often displayed on two graphs: gain versus frequency and phase versus frequency.

In *transient analysis,* a circuit's response as a function of time is computed. Thus you wind up with a graph of voltage versus time.

Let's now turn to a widely used circuit design simulation, and analysis software package that, as its promoters declare, "puts an electronics lab in a computer." Although many excellent schematic and simulation programs exist, here we examine Multisim V6.2, from Electronics Workbench.

Multisim V6.2

To give you a feel for what computer circuit design, simulation, and analysis is like, we look, briefly, at how Multisim builds circuits, works with instruments, and performs circuit simulation and analysis.

Building a Circuit

The user interface for Multisim is shown in Figure 16.3. It consists of the following elements:

- A system toolbar with buttons for commonly-performed functions.
- The Multisim Design Bar that allows you easy access to sophisticated functions offered by the program.
- The "In Use" list that lists all the components used in the current circuit.

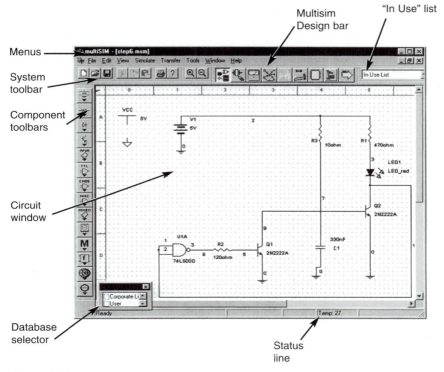

Figure 16.3
Electronics Workbench Multisim user interface.
Source: Electronics Workbench Multisim™

- The component toolbars containing Parts Bin buttons. These buttons let you open component family toolbars which in turn contain buttons for each family of components in the Parts Bin.
- The Circuit Window, where circuit design takes place.
- The database selector that allows you to choose which database levels are to be visible as component toolbars.
- The status line that displays information about the current operation and a description of the item the cursor is currently pointing to.

In designing a circuit, first select and drag various components from the tool bins to your circuit window, or drawing area. (See Figure 16.4.) Next, select component values and reference IDs as needed.

With components in place, it's a simple matter to wire them together. Using a mouse, point to a component's terminal (a short protruding line

Placing the cursor on this component toolbar Parts bin...

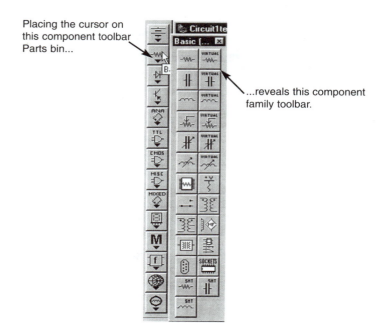

...reveals this component family toolbar.

Figure 16.4
Toolbar Parts Bin.
Source: Electronics Workbench Multisim™

on the component) to highlight it. Then press and hold the mouse button, and drag it so a wire appears. Next, continue to drag the wire to a terminal on another component, and when the new terminal is highlighted, release the mouse button. The wire is automatically routed at a right angle.

It should be noted, with Multisim, you can choose to wire components either automatically or manually. With the former, Multisim automatically wires the connection for you, selecting the best path between your chosen pins. Automatic wiring avoids wiring through other components or overlapping wires. With manual wiring, you control the path of the wire on the circuit window. You can even combine the two methods in a single wire, for example, start wiring manually and then let Multisim automatically complete the wire for you.

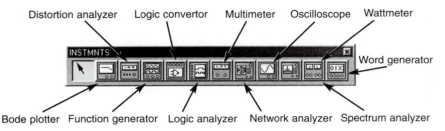

Figure 16.5
Instruments toolbar.
Source: Electronics Workbench Multisim™

Working With Instruments

Multisim includes 11 virtual instruments. (See Figure 16.5.) By bringing up the Instruments button on the Design Bar, you have access to a distortion analyzer, logic converter, multimeter, oscilloscope, wattmeter, word generator, spectrum analyzer, network analyzer, logic analyzer, function generator, and Bode plotter, the last of which is not available in the real world.

With each virtual instrument, you have two views: the instrument icon you attach to your circuit, and the opened instrument, where you set the instrument's controls and display options. (See Figure 16.6.)

Circuit Simulation and Analysis

To simulate a circuit with Multisim, you click the Simulate button in the Design Bar and, from the pop-up menu, choose Run/Stop.

When the simulation begins, you need some way of displaying the results. An oscilloscope will do nicely.

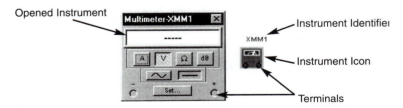

Figure 16.6
Virtual Multimeter.
Source: Electronics Workbench Multisim™

Figure 16.7
Circuit simulation.
Source: Electronics Workbench
Multisim™

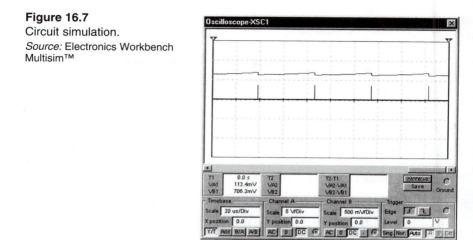

To see the results on the scope, you simply double-click on the oscilloscope icon to "open" the instrument display, if it isn't already open. You will see results similar to what is shown in Figure 16.7.

Multisim provides many different types of analyses. When you perform an analysis, the results are displayed on a plot in the Multisim Grapher (unless you specify otherwise) and saved for use in the Postprocessor. An example of a transient analysis is shown in Figure 16.8.

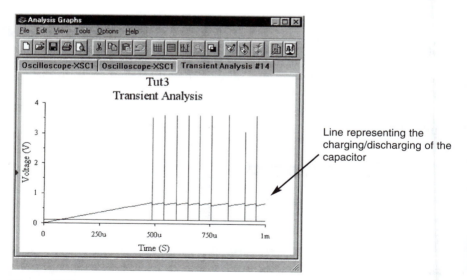

Figure 16.8
Circuit analysis.
Source: Electronics Workbench Multisim™

As we have seen, computer-based circuit design and analysis tools are available to bring you quickly and effectively from concept to breadboarding. As an electronics technician, you will want to investigate them, explore them, and use them to enhance the design process, whenever you can.

Summary

In Chapter 16, we peeked at the world of virtual electronics, where circuit design, simulation, and analysis is done on a computer. We then examined a typical software package, Multisim V6.2, from Electronics Workbench.

Review Questions

1. It is now possible to both _____ a circuit and _____ its operation on a computer.

2. With a virtual circuit, there are _____, _____, and _____ stages.

3. A circuit, real or virtual, consists of electronic _____ connected by _____ (or traces on a PC board).

4. In circuit simulation software, components must be endowed with characteristics in the form of _____ and _____.

5. Circuit simulation calculates a numerical solution to a _____ representation of the circuit you create.

6. Most general-purpose simulations have four main stages: _____, _____, _____, and _____.

7. DC _____ _____ analysis determines the dc operating point of a circuit.

8. In ac _____ analysis, the dc operating point is first calculated to obtain linear, small-signal models for all nonlinear components.

9. In _____ analysis, a circuit's response as a function of time is computed.

10. Multisim includes 11 _____ instruments.

CHAPTER 17

Useful Electronic Projects You Can Build

Objectives

In this chapter you will learn:

- How a Logic Probe project is used to test digital circuits.
- How a Carport Night-Light Controller project uses a photoresistor to turn on a light.
- How a Push-Button Combination Lock project is "undefeatable."
- How a Tone-Burst Generator project continues sounding after a push button is released.
- How an LED Timing Tester project tests your reaction time.
- How a Dual Burglar Alarm project works with both normally open and normally closed contacts.
- How a Telephone Hold Button project puts incoming and outgoing calls on hold.
- How a Logic Pulser project creates a bounceless pulse.
- How a Five- and Nine-Volt Power Supply project provides dc.
- How an LED "Night Rider" project sequences 16 LEDs back and forth.

- How a simple Continuity Tester can be made with a battery, a resistor, an LED, and a buzzer.

- How a DC Motor Driver and Reverser is constructed around an analog comparator.

- How an Attic Fan Actuator automatically turns on an attic fan.

- How an Emergency strobe light operates from a vehicle's 12-volt battery source.

- How a Laser Pointer Remote Control turns on or off a load.

- How the Turn-Signal Reminder alerts you to when a turn signal is stuck in the ON position.

- How to build a simple 2W Stereo Amplifier to operate with your Walkman® or similar device.

By now you're probably eager to get some hands-on electronics experience. I'll bet you want to build, test, and troubleshoot the kind of electronic devices we've been talking about. In this chapter we offer you the chance.

In the next few pages we present twelve electronic projects we hope you'll find both interesting and useful. Furthermore, they're all relatively simple to build. Just pick one that interests you, gather the components, choose a breadboarding approach that's most appropriate (see Chapter 10), and get started.

Yes, you'll make mistakes; the project may not work the first time. But remember: Every time you look at a schematic drawing, pick up a component to identify its characteristics, and make a connection, you are learning electronics by doing electronics. I know your confidence is going to soar as eventually LEDs light, speakers sound, and relays click. Go for it! This is your chance to "make like" an electronics technician.

Logic Probe Project

Schematic

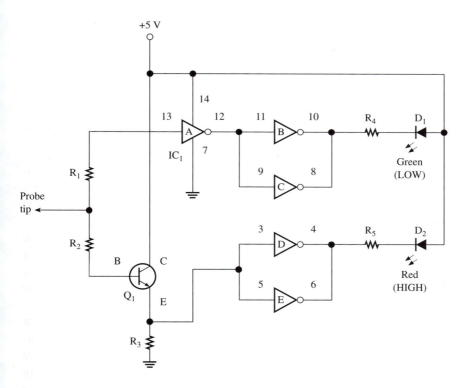

Parts List

D_1	Green Light-Emitting Diode (LED)
D_2	Red Light-Emitting Diode (LED)
IC_1	7404 Hex-Inverter
Q_1	2N3904 NPN Transistor
R_1	1,000-ohm Resistor
R_2	10,000-ohm Resistor
R_3	470-ohm Resistor
R_4, R_5	220-ohm Resistor

Project Description

A logic probe is used to indicate a low, high, or pulse condition in a digital circuit. It gets its power from the circuit under test. When the probe tip (a simple piece of wire in this case) is connected to a low point in the circuit under test, the green LED will light. When it touches a high point in the circuit, the red LED will light. If the point being tested is pulsing, both LEDs will appear on.

Construction Hints

By building the Logic Probe Project at one end of a solderless circuit board, you have a handy logic probe for testing circuits built elsewhere on the board.

Carport Night-Light Controller Project

Schematic

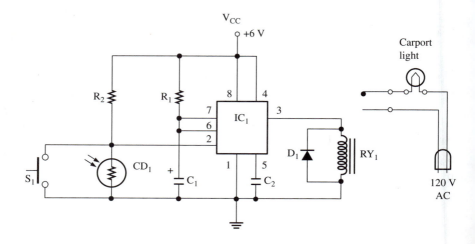

Parts List

C_1	4.7 μF Capacitor
C_2	0.01 μF Capacitor
CD_1	Photoresistor
D_1	1N4001, 1 A/50 V Diode (Rectifier)
IC_1	555 Timer
R_1	2,200,000-ohm Resistor
R_2	10,000-ohm Resistor
RY_1	6-volt Relay
S_1	N.O. Push-button Switch

Project Description

This project will automatically turn on your carport light for a couple of minutes when you drive in at night. Position the photoresistor so that it is in the dark during the day, but is illuminated by your car's headlights as you drive in at night. Switch S_1 can be used to trigger the circuit manually as you leave for the evening.

Construction Hints

The value of R_1 and C_1 determine *on* time of the carport light. Increasing either one increases the *on* time.

Mount S_1 near the door to your house for easy access. Be sure the relay contacts can handle the load current of your carport light.

Push-Button Combination Lock Project

Schematic

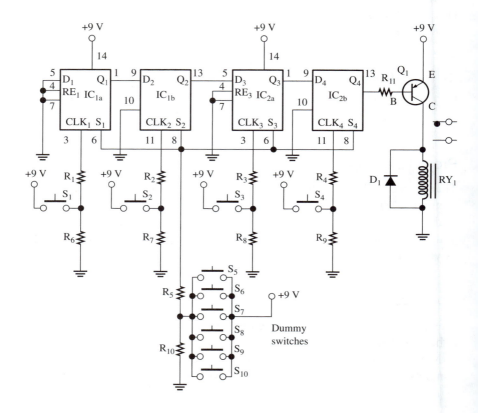

Parts List

D_1	1N4001, 1 A/50 V Diode (Rectifier)
IC_1, IC_2	4013 D Flip-Flop
Q_1	2N3906 PNP Transistor
R_1–R_5	120,000-ohm Resistor
R_6–R_{10}	3,300,000-ohm Resistor
R_{11}	10,000-ohm Resistor
RY_1	6-volt Relay
S_1–S_{10}	N.O. Push-button Switch (or 10-button keypad)

Project Description

Pressing switches S_1 through S_4 in sequence will activate Q_1 and turn on the relay. Pressing any switch S_5 through S_{10} will interrupt the sequence. Thus S_5 through S_{10} are dummy switches. Before activating the relay, press any dummy switch. Then begin the sequence. Use as a lock for power equipment, such as a drill press.

Construction Hints

Mount all ten switches on a panel. Since an unauthorized person does not know which switches are the dummy switches, or, for that matter, that there are active switches which must be pressed in sequence, it is almost impossible to break the code.

Tone-Burst Generator Project

Schematic

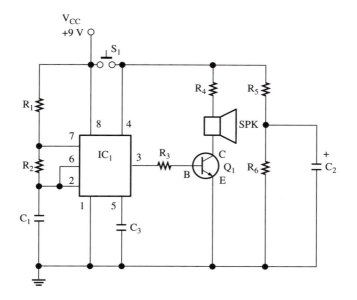

Parts List

C_1	0.1μF Capacitor
C_2	100μF Capacitor
C_3	0.01μF Capacitor
IC_1	555 Timer
Q_1	2N3904 NPN Transistor
R_1	2,200-ohm Resistor
R_2	33,000-ohm Resistor
R_3	220-ohm Resistor
R_4	10-ohm Resistor
R_5	3,300-ohm Resistor
R_6	12,000-ohm Resistor
SPK	8-ohm Speaker
S_1	N.O. Push-button Switch

Project Description

When switch S_1 is pressed, a tone is heard from the speaker. When S_1 is released, the tone continues for a time determined by C_2 and R_6.

Construction Hints

Increasing the value of either C_2 or R_6 increases the time the tone is heard after the switch is released. Conversely, decreasing the value of either component will reduce the *on* time after the switch is released.

LED Timing Tester Project

Schematic

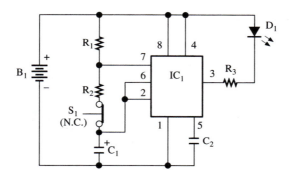

Parts List

B_1	9-volt Battery Clip
C_1	10μF Capacitor
C_2	0.01μF Capacitor
D_1	Red Light-Emitting Diode (LED)
IC_1	555 Timer
R_1	820-ohm Resistor
R_2	47,000-ohm Resistor
R_3	330-ohm Resistor
S_1	N.C. Push-button Switch

Project Description

If S_1 is left closed, the LED flashes for one-tenth of a second approximately every 6 seconds. The object of the game is to anticipate the flash rate. If you press S_1 at exactly the time the LED lights, holding S_1 down will cause the LED to stay lit. If you press S_1 at any other time, the LED will remain off.

Construction Hints

Makes a great attention-getter project. Mount it in a sturdy plastic or metal box and leave it on the coffee table.

Dual Burglar Alarm Project

Schematic

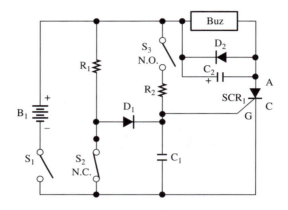

Parts List

B$_1$	9-volt Battery Clip
Buz	6-volt Buzzer
C$_1$	0.1μF Capacitor
C$_2$	25 to 50μF Capacitor
D$_1$, D$_2$	1N4001, 1 A/50 V Diode (Rectifier)
R$_1$, R$_2$	47,000-ohm Resistor
S$_1$	Single-Pole, Single-Through (SPST) Switch
S$_2$	Normally Closed Contact (N.C.)
S$_3$	Normally Open Contact (N.O.)
SCR$_1$	C106B1 Silicon Controlled Rectifier

Project Description

This simple alarm will sound a buzzer when normally open (N.O.) or normally closed (N.C.) contacts are activated. The buzzer will stay on even if the contacts are brought back to their normal position. Thus it is a latching burglar alarm. Switch S$_1$ resets the alarm. A key lock can be substituted for S$_1$.

Construction Hints

To use, connect contact S_2 to any normally closed switch such as security foil on a window. If you do not wish to use this contact, simply connect the leads together. Additional normally closed contacts can be connected in parallel.

Connect S_3 to any normally open switch such as a magnetic door switch. If you do not wish to use this contact, simply leave it open. Additional normally open contacts can be connected in series.

Telephone Hold Button Project

Schematic

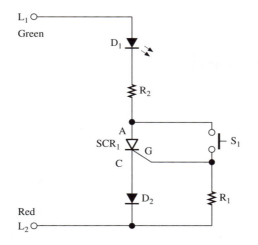

Parts List

D_1	Red Light-Emitting Diode (LED)
D_2	1N4001, 1 A/50 V Diode (Rectifier)
R_1	820-ohm Resistor
R_2	1,200-ohm Resistor
S_1	Normally Open (N.O.) Push-button Switch
SCR_1	C106B1 Silicon Controlled Rectifier

Project Description

With this simple hold button, all incoming and outgoing calls may, at the press of your finger, be placed on hold. A red LED indicates the party is still waiting.

Construction Hints

To install, remove the telephone case and locate the red and green wires. From the green wire make a connection to L_1 and from the red wire a connection to L_2. With some phone systems you may need to reverse these connections.

To test, call a friend. While pressing the push button, hang up the phone. As soon as the phone is on the hook, you may let go of the push button. The LED will remain lit, with your friend on hold.

Logic Pulser Project

Schematic

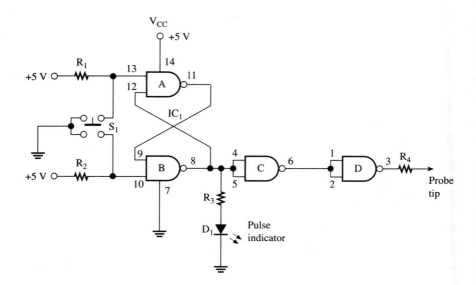

Parts List

D_1	Red Light-Emitting Diode (LED)
IC_1	7400 Quad 2-input NAND Gate
R_1, R_2	820-ohm Resistor
R_3	220-ohm Resistor
R_4	330-ohm Resistor
S_1	Double-Pole, Double-Through Spring-Activated Push-button (DPDT) Switch

Project Description

The logic pulser is a "relative" of the logic probe. With the pulser, we generate a "clean," bounceless pulse (one that creates no "extra" lows and highs as it transitions from one state to another) with just the press of a switch. Such an instrument is handy for producing pulses at various points in a circuit under test.

Construction Hints

As with the Logic Probe Project, you should consider building this project at one end of a solderless circuit board. You will then have a handy logic pulser for testing circuits built elsewhere on the board.

Five- and Nine-Volt Power Supply Project

Schematic

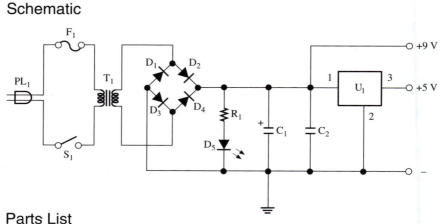

Parts List

C_1	1,000µF Capacitor
C_2	0.1µF Capacitor
D_1–D_4	1N4001, 1 A/50 V Diode (Rectifier)
D_5	Red Light-Emitting Diode (LED)
F_1	1 A Fuse
PL_1	Line Cord
R_1	470-ohm Resistor
S_1	Single-Pole, Single-Through (SPST) Switch
T_1	6.3 V Filament Transformer
U_1	LM7805, 5 V Regulator

Project Description

The purpose of any power supply is to change the 120-volt ac coming from the wall socket to a smooth dc. This power supply has two outputs: a regulated 5 volts and a nonregulated 9 volts. An excellent project for your workbench.

Construction Hints

Since this project operates from the wall socket, extreme caution must be exercised in its use. Be sure to house the Power Supply in an enclosure made of plastic or metal.

LED "Night Rider" Project

Schematic

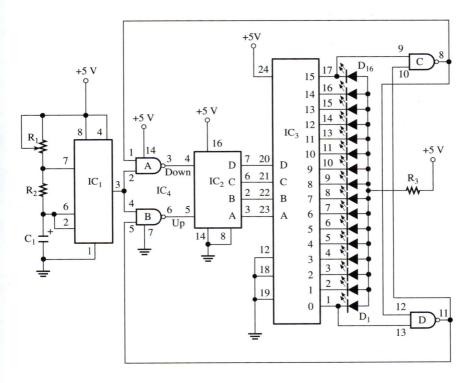

Parts List

C_1	47 μF Capacitor
D_1–D_{16}	Red Light-Emitting Diode (LED)
IC_1	555 Timer
IC_2	74193 Up/Down Binary Counter
IC_3	74154 1-of-16 Decoder
IC_4	7400 Quad, 2-input NAND Gate
R_1	100,000-ohm Potentiometer
R_2	1,000-ohm Resistor
R_3	220-ohm Resistor

Project Description

LEDs light in sequence, back and forth, with a speed determined by the setting of R_1. A great attention getter.

Construction Hints

Mount the LEDs in a three-quarter circle for added effect. Note: this circuit will not run on a 9-volt battery. You must use a 5-volt power supply or a 6-volt battery (four C cells in series will work).

Continuity Tester Project

Schematic

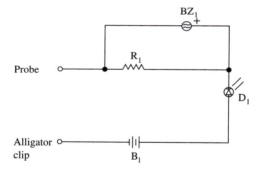

Parts List

D_1	Light-Emitting Diode (LED), any color
B_1	9-Volt Alkaline Battery
R_1	470-ohm Resistor
BZ_1	Piezo-buzzer (Radio Shack 273-060, or equivalent)
Probe	(Radio Shack 278-704 or equivalent)
Clip	Heavy-duty alligator clip (Radio Shack 270-349, or equivalent)

Project Description

Here is a simple continuity tester with an audible and visual output. This is a great tester to use in checking for low resistance shorts and circuit continuity between cables and connectors. It is also handy for checking out switches. Be sure all power is removed from any circuit under test.

Construction Hints

You could build the circuit by wrapping it around the 9-volt battery, literally. Or use perforated board with push-in terminals.

DC Motor Driver and Reverser Project

Schematic

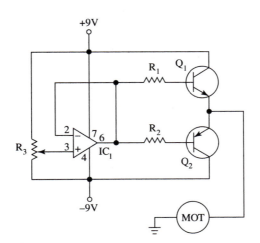

Parts List

IC_1	741 Op-amp
MOT	Small DC Motor
Q_1	TIP3055 Power NPN Transistor (equivalent part is SK3960)
Q_2	MJE34 Power PNP Transistor (equivalent part is SK3274)

R_1, R_2 1,000-ohm

R_3 25,000-ohm Potentiometer

Project Description

The potentiometer, R_3, provides speed and direction control. When the wiper of R_3 is at its mid-position, the op-amp's output will be near zero and neither Q_1 nor Q_2 will turn on. When the wiper is turned up toward the positive side, the output will go positive and Q_1 will supply current to the motor. When R_3 is turned toward the negative supply, the IC's output switches to a negative voltage, turning Q_2 on and Q_1 off. The motor's direction is reversed. As R_3's wiper is moved toward either end, the speed increases in whichever direction it is turning.

Construction Hints

You could build the project on solderless circuit board or on perforated board with push-in terminals.

Attic Fan Actuator

Schematic

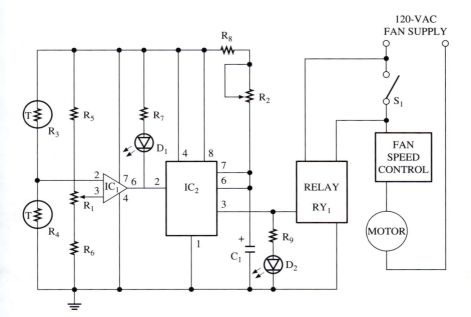

Parts List

C_1	47µF Capacitor
D_1, D_2	Red Light-Emitting Diode (LED)
IC_1	741 Op-Amp
IC_2	555 Timer
R_1	10,000-ohm, 10-turn Potentiometer (threshold control)
R_2	500,000-ohm Potentiometer (fan run time)
R_3, R_4	10,000-ohm Thermistor (Radio Shack 271-110)
R_5, R_6	10,000-ohm Resistor
R_7, R_9	1,200-ohm Resistor
R_8	47,000-ohm Resistor
RY_1	6-volt Relay
S_1	Single-Pole, Single-Through (SPST) Switch

Project Description

The attic fan actuator is a heat-sensitive circuit that automatically senses the difference in attic and eave temperature and turns on attic fans when they can draw in cooler air. The project is operated from a 12-volt dc source, uses two solid-state components, a comparator and a 555 Timer IC, and two thermistors. Depending on the relay chosen, any attic fan motor can be controlled.

Construction Hints

This circuit measures temperature difference, not temperature. Once the difference passes a certain threshold, the 555 IC Timer is triggered, activating the relay.

One thermistor is located in the attic, the other in the eave. Be sure the relay contacts are capable of handling the fan motor's current.

Emergency Strobe Light

Schematic

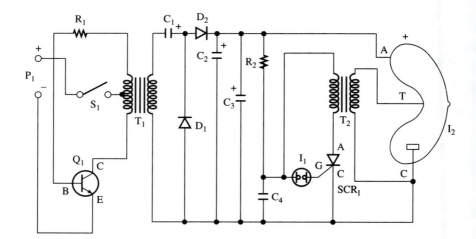

Parts List

C_1	22 μF/350 V Capacitor
C_2, C_3	47 μF/250 V Capacitor
C_4	0.47 μf/250 V Capacitor
D_1, D_2	1N4005, 1A/600V Diode (Rectifier)
I_1	NE-2 Neon Bulb
I_2	5-W Xenon Flash tube
P_1	12-Volt Cigarette Light Plug (Male) Adapter
Q_1	2N3055 Power Transistor
R_1	2,700-ohm Resistor
R_2	1,000,000-ohm Resistor
SCR_1	C106B1 Silicon Controlled Rectifier
S_1	Single-Pole, Single-Through (SPST) Switch
T_1	24 V, 400 mA Center-Tapped Transformer
T_2	4 kV Trigger Transformer

Project Description

The Emergency Strobe Light Project operates from a vehicle's 12-volt battery source. You simply plug it in and place the project in a prominent position on the vehicle's hood, roof, or trunk. You are now assured of being visible to passing vehicles.

Construction Hints

The project consists of three basic circuits: the inverter, the voltage doubling power supply, and the trigger and flash circuit. The Xenon light flashes approximately 84 times per minute and consumes about 650 mA at 12 volts.

All components should be assembled on a printed circuit board and the entire circuit installed in a cabinet. Of course, the Xenon tube must be placed on top of the cabinet. It would be nice if a reflector of some sort were mounted behind the Xenon tube to "aim" the flash in a particular direction.

Laser Pointer Remote Control

Schematic

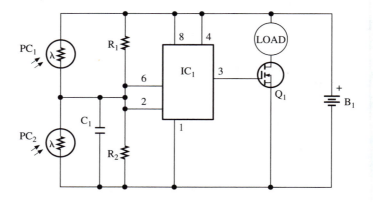

Parts List

B_1	9-volt Battery Clip
C_1,	0.1 μF Capacitor
IC_1	TLC555 Timer
PC_1, PC_2	Photoresistor
Q_i	IRF510
R_1, R_2	10,000-ohm Resistor

Project Description

The Laser Pointer Remote Control Project turns on or off a load when a laser pointer beam is directed at it. When light from the laser pointer hits one of the photocells, the load turns on and stays on. When light from the pointer hits the other photocell, the load turns off and stays off. When the two photocells are receiving the same ambient light, the load stays in whatever state is was already in.

Construction Hints

The two photocells should be mounted as far apart as possible. If they are too close, it will be difficult to accurately aim the laser pointer to trigger them individually.

Turn-Signal Reminder

Schematic

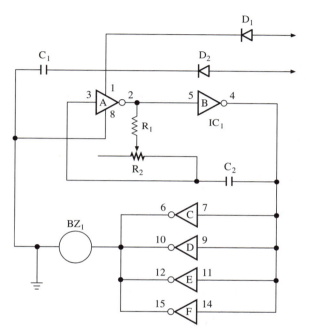

Parts List

BZ_1	Piezo-Buzzeer (Radio Shack 273-060, or equivalent)
C_1	0l μF Capacitor
C_2	0.01 μF Capacitor
D_1, D_2	1N4002, 1 A/100 V Diode (Rectifier
IC_1	CD4049 CMOS Hex Inverter
R_1	47,000-ohm Resistor
R_2	250,000-ohm Potentiometer

Project Description

You've seen them, right ahead of you, those drivers with their vehicle's turn signal stuck in the ON position for miles and miles. While you may not be able to change what other drivers do, by installing this Turn-Signal

Reminder Project in your vehicle, you will be able to avoid being stuck in the same predicament.

The project is built around the popular, inexpensive, readily available CD4049 CMOS hex-inverter-buffer IC. A piezoelectric sounder is the audio source. Power, of course, is derived from the vehicle's battery.

Construction Hints

The anodes of the two diodes, D_1 and D_2, are connected to the right and left turn-signal outputs, which are normally at zero volts when neither signal is activated. Operating either turn signal supplies a pulsing voltage to either diode's anode, thus powering the circuit, causing BZ_1 to emit an audio tone.

Two-Watt Stereo Amplifier

Schematic

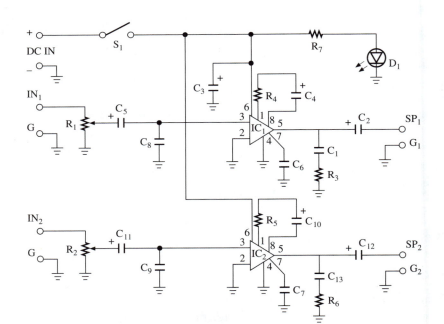

Parts List

C_1, C_{13}	0.047 μF Capacitor
C_2, C_{12}	330 μF/10 V Capacitor
C_3	100 μF/16 V Capacitor
C_4, C_{10}	10 μF/16 V Capacitor
C_5, C_{11}	4.7 μF/50 V Capacitor
C_6, C_7	0.1 μF Capacitor
C_8, C_9	680 pF Capacitor
D_1	Red Light-Emitting Diode (LED)
IC_1, IC_2	LM286 Amplifier
R_1, R_2	10,000-Potentiomete
R_3, R_6	10-ohm Resistor
R_4, R_5	1,200-ohm Resistor
R_7	470-ohm Resistor
S_1	Single-Pole, Single-Through (SPST) Switch

Project Description

This is a relatively simple, portable, easy to use stereo amplifier that plugs into the headphone jack of a Walkman® or similar device. Operated from four AA cells, the project includes two identical LM386 integrated circuit amplifier chips, one for each stereo channel. Separate volume adjustment for each channel is provided via a potentiometer.

Construction Hints

It is best to construct your project on a printed circuit board, then mount the entire assembly in a sturdy case.

To get the best results from the stereo amplifier, adjust the volume control on your Walkman®, or similar device, so that the sound is loud, yet without distortion, when the variable potentiometers are in their middle position.

Summary

In Chapter 17, we examined the operation and construction of 17 simple, easy to build electronic projects, while gaining experience in reading schematic drawings.

Review Questions

1. A _____ _____ is used to indicate a low, high, or pulse condition in digital circuit.

2. A photoresistor is sensitive to _____.

3. In the Push-Button Combination Lock project, pressing switches _____ through _____ in sequence will activate Q_1.

4. In the Tone-Burst Generator project, the tone produced continues for a time determined by the values of _____ and _____.

5. In the LED Timing Tester project, if S_1 is left closed, the LED flashes for one-tenth of a second approximately every _____ seconds.

6. In the Dual Burglar Alarm project, the buzzer will sound if the normally open contacts are _____ or if the normally closed contacts are _____.

7. In the Telephone Hold Button project, a red LED indicates the party is still _____.

8. A Logic Pulser project generates a clean, _____ pulse.

9. In the Five- and Nine-Volt Power Supply project, the five-volt output is _____.

10. In the LED "Night Rider" project, the back and forth speed is determined by the setting of _____.

11. The Continuity Tester project gives an _____ and _____ indication of continuity.

12. With the DC Motor Driver and Reverser, both speed and direction control are provided by a _____.

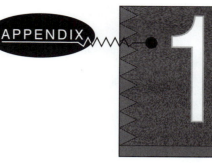

Electronic Kit and Part Sources

The following represents a list, in alphabetical order, of many USA sources of electronic kits. Some are manufacturers, some are vendors, some are both. Contact them via phone, fax, email, or through their web site.

(List supplied courtesy of Fred Bleckman, author of *Simple Low-Cost Electronics Projects*, 1998, ISBN 1-878707-46-9.)

Alltronics
2300-D Zanker Road
San Jose, CA 95131
Phone: (408) 943-9773
Fax: (408) 943-9776
Web site: www.alltronics.com/kits.htm
Email: ejohnson@alltronics.com

Cal West Supply, Inc. (Hallbar electronic kits)
3835 R. East Thousand Oaks Blvd. #204
Westlake Village, CA 91362
Phone: (800) 892-8000 or (805) 497-9900
Fax: (805) 557-0249
Web site: www.hallbar.com
E-mail: hallbar@hallbar.com

Carl's Electronics
P.O. Box 182
Sterling, MA 01564
Phone: (978) 422-5142
Fax: (978) 422-8574
Web site: www.electronickits.com
Email: sales@electronickits.com

C & S Sales (Elenco kits)
150 West Carpenter Ave.
Wheeling, IL 60090
Phone: (800) 292-7711 or (847) 541-0710
Fax: (847) 541-9904
Web site: www.cs-sales.com
E-mail: info@cs-sales.com

Circuit Specialists, Inc.
220 S. Country Club Dr.
Mesa, AZ 85210
Phone: (800) 528-1417 or (480) 464-2485
Fax: (480) 464-5824
Web site: www.web-tronics.com
E-mail: jruss@cir.com

Digi-Key Corporation
701 Brooks Avenue South
Thief River Falls, MN 56701
Phone: (800) 344-4539
Fax: (218) 681-3380
Web site: www.kigikey.com

Digital Products Co.
134 Windstar Circle
Folsom, CA 95630
Phone: (916) 985-7219
Fax: (916) 985-8460

Earth Computer Technologies
1110 Cale Cordillera
San Clemente, CA 92673
Phone: (949) 361-2333
Fax: (949) 361-2121
Web site: www.flat-panel.com

EKI (Electronics Kits International, Inc.)
P.O. Box 970431
Orem, UT 84097-0431
Phone: (800) 453-1708
Fax: (801) 224-5880
Web site: www.eki.com
E-mail: emilio@eki.com

Electronic Goldmine
P.O. Box 5408
Scottsdale, AZ 85261
Phone: (800) 450-0697
Fax: (480) 661-8259
Web site: www.goldmine-elec.com
Email: goldmine-elec@goldmine-elec.com

Electronic Rainbow, Inc.
6227 Coffman Road
Indianapolis, IN 46268
Phone: (888) 291-7262
Fax: (317) 291-7269
Web site: www.labvolt.com

ElectronicsUSA.com
14270 Apple Creek Dr.
Victorville, CA 92392
Phone: (760) 241-5775 or (775) 416-8075
Web site: www.electronicsusa.com
Email: info@electronicsusa.com

Electronix Express
365 Blair Road
Avenal, NJ 07001
Phone: (732) 381-8020
Fax: (732) 381-1006
Web site: www.elexp.com
Email: electron@elexp.com

EMAC, Inc.
11 Emac Way
Carbondale, IL 62901
Phone: (618) 529-4525
Fax: (618) 457-0110
Web site: http://www.emacinc.com
Email: info@emacinc.com

Gateway Electronics, Inc.
8123 Page Blvd.
St. Louis, MO 63130
Phone: (800) 669-5810 or (314) 427-6116
Fax: (314) 427-3147
Web site: www.gatewayelex.com
Email: gateway@mvp.net

Graymark International, Inc.
Box 2015
Tustin, CA 92781
Phone: (800) 854-7393
Fax: (714) 544-2323
Web site: www.graymarkint.com
Email: sales@graymarkint.com

HobbyTron.com
1185 South 1480 West
Orem, UT 84058
Phone: (800) 422-1100 or (877) 606-8766
Fax: (800) 470-1606
Web site: www.hobbytron.com
Email: brad@littlefishcommerce.com

Information Unlimited
P.O. Box 716
Amherst, NH 03031
Phone: (800) 221-1705 or (603) 673-4730
Fax: (603) 672-5406

LNS Technologies
Box 67243
Scotts Valley, CA 95067
Phone: (831) 438-2028
Fax: (831) 438-0661
Web site: www.techkits.com
Email: Instech@ncal.verio.com

Marcraft International Corp.
100 N. Morain St.
Kennewick, WA 99336
Phone: (800) 441-6006
Fax: (509) 374-9250
Web site: www.mic-inc.com
Email: mcraft@oneworld.owt.com

Marlin P. Jones & Associates, Inc.
P.O. Box 12685
Lake Park, FL 33403
Phone: (800) 652-6733
Fax: (800) 432-9937
Web site: www.MPJA.com
Email: mpja@mpja.com

Miller Engineering
P. O. Box 282
New Canaan, CT 06840
Phone: (203) 595-0619
Web site: www.microstru.com

Namco, Inc.
P. O. Box 9-102
Hwy. 81N.
Calhoun, KY 42327
Web site: nimcoinc.com

Parallax, Inc.
599 Menlo Dr. #100
Rocklin, CA 95765
Phone: (888) 512-1024 or (619) 624-8333
Fax: (916) 624-8003
Web site: www.parallaxinc.com

Quality KIts
49 McMichael St.
Kingston, Ontario
Canada KLMlm8
Phone: (888) 464-5487 or (613) 544-6333
Fax: (613) 544-4944
Web site: www.qkits.com
Email: anything@qkits.com

Radio Shack Product Support
200 Taylor St., Suite 600
Fort Worth, TX 76102
Phone: (800) 843-7422
Fax: (817) 415-2303
Web site: www.radioshack.com

Ramsey Electronics, Inc.
793 Canning Parkway
Victor, NY 14564
Phone: (800) 446-2295 or (716) 924-4560
Fax: (716) 924-4886
Web site: www.ramseyelectronics.com
Email: OrderDesk@ramseyelectronics.com

Robotikitsdirect Company
17141 Kingsview Ave.
Carson, CA 90746
Phone: (310) 515-6800
Web site: www.owirobot.com
Email: robotikitsdirect@pacbell.com

Ten-Tec, Inc.
1185 Dolly Parton Pkwy.
Sevierville, TN 37862
Phone: (800) 833-7373 or (865) 453-7172
Fax: (865) 428-4483
Web site: www.tentec.com
Email: sales@Tentec.com

Transtronics
3209 W. 9th St.
Lawrence, KS 66049
Phone: (785) 841-3089
Fax: (785) 841-0434
Web site: www.xtronics.com/kits.htm
Email: kits@xtronics.com

USI Corp.
P.O. Box N2052
Melbourne, FL 32902
Phone: (321) 725-1000
Fax: (321) 723-6784

Weeder Technologies
P.O. Box 2426
Fort Walton Beach, FL 32549
Phone: (850) 863-5723
Fax: (850) 863-5723
Web site: www.weedtech.com
Email: tenny@weedtech.com

Velleman Inc.
7415 Whitehall St., Suite 117
Fort Worth, TX 76118
Phone: (817) 284-7785
Fax: (817) 284-7712
Web site: www.velleman.be
Email: velleman@earthlink.be

Worldwyde
33523 Eight Mile Road, #A3-261
Livonia, MI 48152
Phone: (800) 773-6698
Fax: (248) 474-0605
Web site: www.worldwyde.com
Email: sales @worldwyde.com

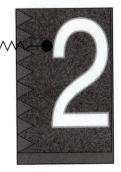

The Big Three
Electronics Magazines

Poptronics

This magazine represents the fusing of two magazines well known in the field for decades: *Popular Electronics* and *Electronics Now*. Here are some observations:

- It is strong on beginner construction articles. It is not embarrassed to print some easy projects.
- The magazine is well laid out, with excellent text and graphics—it is a pleasure to hold and read.
- It is easy to read; the vocabulary is suited to the average reading level of an electronics technician student.

Micro Times

While there are dozens of computer magazines available, *Micro Times* is not only good—it is free. Filled with short, snappy articles on everything from e-commerce to mobile computing, it is ideal for those wanting to keep up on technical things and on the business end of the industry. Here are some observations:

- It has a large page format.
- There are many excellent columns that appear regularly.
- The magazine is loaded with product ads that let you know what is currently available and at what price.

Nuts and Volts Magazine

An excellent project magazine that's also loaded with theory you can understand. Here are some observations:

- It has an excellent presentation in a large-page format.
- There is good coverage of microcontrollers.
- The magazine always has unusual articles on everything from lasers to electric cars.

Glossary

accreditation   For schools or college officially recognized as meeting the essential requirements for academic excellence.

advanced electronics courses   Electronics courses, usually taken in the second year of a 2-year program, that provide specialized knowledge. May be of an elective nature.

apprentice   A person who works for another in order to learn a trade.

associate's degree   A degree granted by a community college, junior college, and some 2-year proprietary schools after completion of a 2-year course of study.

automotive electronics   A subfield of electronics dealing with electrical and electronic systems in motor vehicles.

avionics   A branch of electronics concerned with electronic applications in aviation.

bachelor's degree   A degree awarded by a college or university for completion of an undergraduate program, usually lasting 4 years.

breadboarding   The purpose of constructing a temporary circuit for the purpose of testing and/or modifying a design.

certificate   A document attesting to the fact that a person has completed an educational course of study. At a community college, usually courses in the major only.

communications electronics A large subfield of electronics that includes audio, video, and data telecommunications.

community college A nonresidential 2-year junior college established to serve a specific community and typically supported in part by local government funds.

co-op education An educational program consisting of both classroom study and on-the-job training.

computer-assisted instruction (CAI) Instruction, carried out on a computer, that can aid in circuit design and understanding.

computer electronics A subfield of electronics dealing with computers, from pocket PCs to mainframes.

consumer electronics A broad subfield of electronics dealing with personal-use items for entertainment, safety, and information.

core electronics courses Electronics courses, usually taken in the first year of a 2-year program, that are required for everyone in the major. They are designed to provide a firm foundation in basic electricity and electronics.

correspondence school A school operating on a system in which study materials and tests are mailed to the students, who in turn mail their work back to the school for grading.

cover letter A letter sent out with a résumé that tells the individual receiving the resume why it is being sent.

electrical age A time span, begun in 1801, but reaching its apex in the last quarter of the nineteenth century, in which electrical devices, operating on electricity (current flowing in wires), spread to home, office, and industry.

electrical engineer One who knows electrical power generation and distribution engineering, and is well versed in generator, motor, transformer, transmission line, and high-voltage technology.

electronics age   A time span, begun with the development of the vacuum tube in 1907, during which control of the electron occurred in a vacuum, with a vacuum tube. Today, such control occurs with solid-state components, such as transistors and integrated circuits.

electronics assembly   Involves the putting together of electronic subunits, consisting of electronic components and circuit boards, to create a final, packaged product.

electronics engineer   One who knows circuit design, logic design, communications engineering, microwave circuitry, control systems, etc.

electronics technician   One who typically installs, tests, repairs, and maintains electronic equipment.

electronics technologist   Usually an individual with a bachelor's degree in applied technology or a related field. The technologist is concerned primarily with applications, unlike the engineer who's work is theory-based.

financial aid   Monies made available by federal and state governments and private sources in the form of grants, loans, scholarships, and employment.

general-education courses   Courses of a general nature, such as those in the natural, social, and behavioral sciences, humanities, language arts, and health and physical education, required to complete an associate's or bachelor's degree.

home study   Instruction in a subject given by mail or over the Internet.

information interview   An interview, the purpose of which is to gather knowledge about a particular career or company as opposed to a job interview where one is seeking employment.

industrial electronics   A subfield of electronics encompassing electronic equipment found on the factory floor, the engineering design laboratory, and, in some cases, the front office.

insertion mount technology (IMT)   The mounting of electronic components so that their leads protrude through holes in a printed circuit board.

instrumentation electronics   A subfield of electronics that deals with the design, repair, and maintenance of sophisticated test equipment, from multimeters to spectrum analyzers, to virtual instruments.

Internet   A network of many networks that interconnect worldwide and use the Internet Protocol (IP).

internship   An official or formal program to provide practical, on-the-job experience for beginners in an occupation.

job shadowing   A way to gain exposure to a work environment, usually as a one day activity, by following, or "shadowing" various employees to see what it is they do.

medical electronics   A subfield of electronics concerned with electronic equipment used in the research, diagnosis, and treatment of diseases.

military electronics   Electronics dealing with military equipment such as smart weapons, shipboard communications, satellite surveillance, and the like.

multimedia   The presentation of information on a computer using graphics, sound, animation, and text.

power supply   An electronic circuit that changes alternating current into direct current.

radio and television broadcast electronics   A subfield of radio and television broadcasting dealing with the installation, repair, maintenance, and operation of electronic equipment in live commercial radio and television broadcasts.

resume   A brief, preferably 1- or 2-page, written account of personal, educational, and professional qualifications and experience, as that prepared by an applicant for a job.

rework   Work done to a circuit to bring it into complete conformance with its original specifications.

schematic diagram   A diagram or drawing that shows the scheme of a circuit. On a schematic diagram, components are represented by graphical symbols.

solder   An alloy of lead and tin used to fuse wires and electronic component leads.

supervisor   One who monitors others while they do tasks in which he or she has a demonstrated expertise.

surface mount technology (SMT)   A revolutionary electronic technology that uses surface-mounted components placed on top of a printed circuit board. Tremendous saving in space is realized with this technology.

tech-prep   A program that links high school students, community college students, and, in many cases, bachelor's degree granting institutions through a 2+2+2 arrangement of non-repeating course work.

time management   In its simplest form, a method of recording the things you need to do, prioritizing the list, and providing a check-off procedure to identify when each item is completed.

trade shows   Association- or organization-sponsored product shows consisting of vendor displays along with seminars and technical sessions.

transducer   A device that uses one form of energy to control another form of energy.

troubleshooting   To look for, locate, and repair problems in electronic equipment.

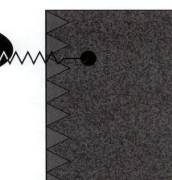

Index